Waste Policy

## NEW HORIZONS IN ENVIRONMENTAL AND ENERGY LAW

**Series Editors:** Kurt Deketelaere, *Professor of Law, University of Leuven, Belgium and University of Dundee, Scotland* and Zen Makuch, *Reader in Law, Barrister, Imperial College, London, UK*

Environmental law – including the pressing considerations of energy law and climate change – is an increasingly important area of legal research and practice. Given the growing interdependence of global society and the significant steps being made towards environmental protection and energy efficiency, there are few people untouched by environmental and energy lawmaking processes.

At the same time, environmental and energy law is at a crossroads. The command and control methodology that evolved in the 1960s and 1970s for air, land and water protection may have reached the limit of its environmental protection achievements. New life needs to be injected into our environmental protection regimes – perhaps through the concept of sustainability in its environmental, economic and social forms. The same goes for energy policy and law, where liberalization, environmental protection and security of supply are at the centre of attention. This important series seeks to press forward the boundaries of environmental and energy law through innovative research into environmental and energy law, doctrine and case law. Adopting a wide interpretation of environmental and energy law, it includes contributions from both leading and emerging international scholars.

Titles in the series include:

EU Environmental Legislation
Legal Perspectives on Regulatory Strategies
*Edited by Marjan Peeters and Rosa Uylenburg*

Polar Oceans Governance in an Era of Environmental Change
*Edited by Tim Stephens and David L. VanderZwaag*

The Fragmentation of Global Climate Governance
Consequences and Management of Regime Interactions
*Harro van Asselt*

Renewable Energy Law in the EU
Legal Perspectives on Bottom-up Approaches
*Edited by Marjan Peeters and Thomas Schomerus*

Environmental Enforcement Networks
Concepts, Implementation and Effectiveness
*Edited by Michael Faure, Peter De Smedt and An Stas*

Regulation of the Upstream Petroleum Sector
A Comparative Study of Licensing and Concession Systems
*Edited by Tina Hunter*

Regional Environmental Law
Transregional Comparative Lessons in Pursuit of Sustainable Development
*Edited by Werner Scholtz and Jonathan Verschuuren*

Earth Governance
Trusteeship of the Global Commons
*Klaus Bosselmann*

Waste Policy
International Regulation, Comparative and Contextual Perspectives
*Alexander Gillespie*

# Waste Policy

## International Regulation, Comparative and Contextual Perspectives

Alexander Gillespie

*Pro-Vice Chancellor of Research and Professor of Law, University of Waikato, New Zealand*

NEW HORIZONS IN ENVIRONMENTAL AND ENERGY LAW

Edward Elgar
PUBLISHING

Cheltenham, UK • Northampton, MA, USA

Published by
Edward Elgar Publishing Limited
The Lypiatts
15 Lansdown Road
Cheltenham
Glos GL50 2JA
UK

Edward Elgar Publishing, Inc.
William Pratt House
9 Dewey Court
Northampton
Massachusetts 01060
USA

A catalogue record for this book
is available from the British Library

Library of Congress Control Number: 2015935897

This book is available electronically in the **Elgar**online
Law subject collection
DOI 10.4337/9781784715885

MIX
Paper from responsible sources
FSC® C013056

ISBN 978 1 78471 587 8 (cased)
ISBN 978 1 78471 588 5 (eBook)

Typeset by Columns Design XML Ltd, Reading
Printed and bound in Great Britain by T.J. International Ltd, Padstow

This book is dedicated to my son Alexander James, for at a very early age, he tried to fathom why the recyclable glass coke bottle had turned into plastic

# Contents

# 1. Introduction

The genesis of this book was nearly 30 years ago when I had a part-time job on a rubbish truck, gathering all types of waste from a middle-class neighbourhood and disposing of it at a local landfill. I got the job as the rubbish business was expanding and more ‘runners’ were required, as greater amounts of garbage were being generated in the community. At the time, I was struck not by the fact that the owner of the business had made a fortune in waste management, but by the realisation that the trash we collected was dumped into landfill that spilled into the harbour. It spilled into the harbour because after having taken waste for more than five decades, the site was hopelessly over-packed, poorly designed and lacking in pollution controls. The recycling, which was meant to be achieved to divert waste from landfill, was a ruse, with large sections of the glass and paper being dumped along with all sorts of other waste, ranging from old food to unwanted consumer appliances. Very little was being sorted, despite the promises made to the well-intentioned community who had a large degree of trust that their waste was being safely and correctly disposed of. In fact, the only effective recycling being done was by a clever entrepreneur who was coming into the dump each week to pick up the large amounts of copper that the Navy was disposing of on a regular basis. It was at this point, as I was musing how this problem looked when multiplied to the global level, that I decided I would write something about the problem. However, in the intervening years, my professional work on environmental matters took me down different routes, and I came to write and engage on topics of conservation and pollution. Now is the time to write the book on waste I always wished to pen.

In the intervening years between my time on the rubbish truck and now, the problem of waste at the site where I used to work has improved greatly. It is no longer a dump. Now, it is a ‘transfer’ station. A wealthy community in the area is conscious of its environmental choices and is proud of its credentials. Inorganic, organic, consumer and hazardous waste are all now well sorted. What can be recycled is dealt with appropriately, and what cannot be recycled is sent to correct destinations that can better handle the waste. Exactly where these places are is far from certain, as New Zealand does not invoke extended producer responsibility schemes, preferring to leave most recycling to the market, not the original producers of the waste. These considerations are now overlaid with a complex regime for the export, or import, of waste, hazardous or not. The end result is that what started out as a local problem, and has been relatively well dealt with by a well-heeled community, has become an international issue. I consider this example to be one small drop in a very large ocean, with the difference being that in my instance where the local situation has improved, the global situation has worsened, in terms of both the amounts and types of waste generated, and the inadequacy of its disposal.

The good news is that in many of the areas of waste disposal, some viable answers are becoming apparent in many – but not all – of the areas. Highlighting what these answers are is an important purpose of this book, in the hope that it can be seen that although serious problems exist, it is possible that many of these difficulties can be adequately dealt with. This is especially so when the philosophies, policies and laws around waste can be seen from a wider, comparative and international perspective – and not in isolation.

The starting point for this study is actual waste and what certainties are known. The first certainty is the definition. Waste is material that is unwanted and has no further use. Humanity has been dealing with waste for thousands of years. Where the certainty has changed is in the difference in the 21st century compared to the past, in that both the types of waste being generated in all countries have multiplied, as have the amounts. We seem to live in a world where waste is omnipresent, and evolving. Our understanding of what makes something hazardous is infinitely more advanced than 100 years ago. Where once there was one hazardous waste, today there are at least 14 hazards to be aware of.

Even when types of waste are not obviously hazardous, the long term implications of it are often still only just being understood. A good current example of this is with plastic. Thirty years ago, all of the drinks I bought to take away came in glass. Today, most things are plastic. I can still recall my fascination at seeing the first plastic Coke bottle, and when finding out that it had no take-back fee attached to it, getting a cold vision of the future. Yet at that time, I could not imagine that the real risk of the plastic tidal wave was not these large plastic items, but the trillions of tiny plastic balls that result from their dissolution and float into the oceans. Nor did it occur to me in my most creative moments that we could actually put so much junk in space, from complete abandoned satellites down to the lost nuts and bolts that held them together, that we would also have a problem of waste in the stratosphere.[1]

I expect in decades to come that there will be new waste streams which I cannot imagine now. I shall leave these for future scholars to study. For this scholar, the waste streams in focus are the ones which are currently most dominant. There are five of these. These are all discussed in Chapter 2, which deals with both definitions and quantities. These are the waste streams which come from the human body; wastes from everyday life which are common; wastes which are hazardous; and wastes which are nuclear or industrial. All of these waste streams are expanding in output. The expansions are linked to growing human population, increasing personal consumption patterns and goods that need to be disposed of becoming economically cheaper to purchase. Even the most highly developed countries, with the best possible recycling schemes and the smartest design systems aiming to reduce waste, find that overall waste production is continuing to grow, albeit at a smaller rate than elsewhere.

The other considerations I think certainty can be held on, also as discussed in Chapter 2, is that few people, and fewer communities, like waste to be near them. Local, national and international policy makers all agree that we should all produce less waste. There are strong environmental and economic reasons for this policy. Socially, it also makes good sense as waste disposal always follows the cheapest possible economic routes. In practice, what this means is that if waste cannot easily be disposed of in the commons for free, such as the oceans or the atmosphere, it will migrate towards the communities which are, in economic terms, less likely to object to its placement. This migration is both a national and international phenomena. Unfortunately, this is the third certainty I believe to be true with waste.

Given these certainties – that waste is increasing, and this is not a desirable thing – the question is how is the problem to be confronted? This question is the second goal of this book. The generic answer of dealing with the problem of waste is trying to approach the issue through what is known as the 'the waste hierarchy'. The waste hierarchy is the focus of Chapter 3. The two aims of the hierarchy are to generate as little waste as possible in the first place, and then to extract the maximum practical benefit of the waste that is still produced. In terms of sequence, the hierarchy suggests that as a priority order in waste prevention and

[1] Rutkin, A. (2014). 'Gone Junk Fishing'. *New Scientist*. 25 Jan. 19; Anon. (2009). 'Space Junk'. *Current Events*. 2 Mar. 4–5.

management legislation and policy, policy makers should pursue prevention; reuse; recycling; other recovery (such as energy recovery); and then – as a last resort – disposal.

The following purpose of this book, once the contextual and theoretical 'answer' to the problem had been found, was to examine how that answer is working in practice. This required an examination of all of the parts of the waste hierarchy, especially in terms of prevention and recycling.

In terms of prevention of waste in the first place, four options are available. First, the product can be prohibited. Second, consumers may voluntarily decide to steer away from the particular stream of waste, or third, pricing mechanisms are placed around it to make it economically unattractive, and thus create an incentive not to create the waste. Finally, the producers of the product that ends up as waste can be brought into the equation via ideals of extended producer responsibility, whereby they become responsible for dealing with the waste item that needs to be disposed of. This last option is growing in popularity and is utilised in over 25 countries, for a multitude of products ranging from consumer batteries to motor vehicles. Through such schemes of extended responsibility, those who control the design of such products are now being given a direct incentive to reduce the waste associated with the product, as they are now responsible for that too.

Recycling and related practices are the top of the second part of the waste hierarchy. This is the topic of Chapter 4. The recycling symbol is one of the most easy to identify labels of environmental responsibility. In the 21st century, following a stop-start love affair with recycling going back over 200 years, it appears that environmental, economic and social goals have finally combined in a more concrete way to support recycling and make this practice the common form of environmental action for many, if not most, citizens. It is also actively pursued by many enterprises, and even governments which are trying to tackle unwanted materials. The goals in this area are well entrenched in domestic, regional and international law and policy, with a number of specific international laws, including the 1989 Basel Convention on the Control of Transboundary Movements of Hazardous Wastes and Their Disposal in particular, being directly targeted to address specific waste streams for recycling and to try to control the trade in materials for recycling which may be hazardous if poorly dealt with. The difficulties in this area are in the limits of this market due to economic concerns and its needs for stimulation, and conversely, when economic returns are great, and the market needs to be controlled. This is a particularly important concern as the quest for profit in this area muddies the distinction between what should be treated as waste and dealt with in one way, and what can be recycled and dealt with in another.

Chapters 5, 6 and 7 proceed down the waste hierarchy, and seek to critically examine the ways in which waste which cannot be prevented at source, or afterwards be recycled, may be disposed. The three disposal options are to burn it, dump it into the ground or dispose of it into the ocean. Dealing with the first option, incineration, is the topic of Chapter 5. The barely controlled burning of waste in open spaces is a practice which goes back thousands of years. Today, it remains a common form of waste disposal in the developing world. Conversely, it is uncommon in developed countries, and especially in urban areas, to light fires to burn waste in an unregulated manner. This practice has largely come to an end because of the clear environmental and social risks of poor incineration and the associated air pollution that is imposed upon surrounding communities when waste is burnt in an ineffective way. These hazards have increased as the waste streams have changed, and the waste that is burnt is more likely to include material with a much greater chance of containing toxic contaminants. It was in large part due to such concerns that the international community stopped this practice of waste disposal on the world's oceans. Thereafter, all incineration has to be done on land. The developed world continues to burn rubbish, but unlike in the past, the standards for the incineration have increased rapidly. The end result is that in many instances, where the highest possible safety and technical standards are applied, the incineration of waste does not cause the same environmental problems that it did in earlier decades. These successes have been supplemented with another benefit in this area, namely energy recovery. This is an issue which evolved out of the 1990s when it became argued that burning waste to dispose of it and make energy from it at the same time may be the best option. However, this potential success for all sorts of incineration only exists for the countries which possess high standards and the state of the art technology to achieve them. In countries without these standards, it is possible that the inefficient burning of waste is a primary source of toxic air pollution.

Chapter 6 is concerned with the disposal of waste into the land. The waste that this chapter covers ranges from that which comes out of the household to that which industry produces. When I wrote this chapter, I was thinking of traditional landfill sites. Collecting and dumping waste into a site of land is the oldest type of waste disposal method known to humanity. To this day, it remains the most common and practiced type of dealing with waste in both developed and developing countries. Given that this is the dominant location for the disposal of waste, an accompanying large body of rules and principles have evolved to manage the problems in this area over the last 100 years. These rules have grown from zoning requirements in the 19th century, to detailed proscription of how to manage these sites, from what can be deposited in dumps and the pollution protection mechanisms to stop leachate, in the 20th century. In the 21st century, these rules and principles have become well adopted and practiced in most developed countries. However, the same cannot be said for many developing countries. This is a particular problem, for many of these countries are now dealing with their own rapidly expanding waste streams, as well as large amounts of imported material which is either intended to be disposed of as waste or, via poor recycling standards, ends up as waste to be deposited into landfill.

The disposal of waste into the ocean is the subject of Chapter 7. Disposing of waste into the ocean is an area where great progress has been made in recent decades. Historically, when it was impractical to incinerate or dump waste into the land, the ocean was always the preferred target for waste disposal. Whilst domestic restraints in this area began in the early part of the 20th century, international ones did not emerge until the end of the 20th century. The international restraints, as found through the 1972 Convention on the Prevention of Marine Pollution by Dumping of Waste and Other Matter (the London Dumping Convention), were the subject of intense debate for over a decade, as the regime prohibited the dumping of only the most hazardous of substances. This approach was completely turned around in 1996, when a protocol to that Convention reversed the assumption that it was safe to dump anything into the ocean and only a few selected waste streams could be deposited into the seas. With this change, a number of waste streams, such as all levels of nuclear waste and all types of industrial waste, were prohibited from discharge into the ocean. In addition, although a number of other waste streams could still be dumped in the ocean, such as dredgings and/or sewage sludge, a much greater regulatory process, designed to make this much safer, became the norm.

The final part of this book, Chapters 8 and 9, go beyond the above considerations to look at two case studies of waste that warrant additional attention. The first, in Chapter 8, covers the topic of the human waste stream. I wanted to focus on this waste as this particular type of waste causes more deaths than all of the other waste streams put together. Despite these impacts, and the concerted efforts of scientists and policy makers to effectively deal with this problem for over 150 years, the progress has become much divided between the developed and the developing world, and where each region is with regard to dealing with human waste as a first, second or third generation type of problem.

The first generation problem was about getting the waste out of human communities. The second generation has been about treating it, and the third, final disposal. Despite strong international interest, the success in achieving the primary goal is mixed, with strong divisions being evident between developed and developing countries. The same conclusion applies for dealing with the international control of disposing treated human waste into freshwater systems. This is not the case for the disposal of treated human waste into the ocean, where the international regulation is much more robust, and as seen in Chapter 7, a trend is developing whereby this practice is becoming less popular. Conversely, the options that are becoming popular for final disposal of sludge are landfill, incineration or recycling. Of all these three, each has both limits and benefits, although the last option, involving the recycling of biosolids into a potential resource, appears to be the best use of the waste stream.

The last chapter in this book, Chapter 9, is about nuclear waste. The reason that this waste stream has a chapter specifically dedicated to it is because this is the most hazardous of all waste types. Ironically, when this waste first appeared, it was not even on the radar as a potential problem. However, within a dozen years, this was the very first type of waste to be specifically focused upon by international law. By the end of the century, once it had become apparent that this was likely to be the last waste left on the planet, tens of thousands of years from now, it became a dominant international concern. Despite this concern and over four decades of debate, the only international rule that is certain is that it should not be dumped into the ocean. Rather, the theory is that it should be reprocessed and/or then, with any residue, set into deep geological storage where it will be held in perpetuity. There are, however, significant problems with both of these options.

# 2. Five certainties on waste

## 1. INTRODUCTION

There are five certainties on waste. The first is what it is, that is, waste involves solid material (that may be liquid, but should be heavier than air) that is unwanted and has no further use. The second certainty is that as the disposal of waste increases, the trend is to find the cheapest way to dispose of it. Often, this method involves it being placed near poor communities. Following on from the recognition that waste often involves social, economic and environmental costs, the third certainty arises – that the trade in waste has often become international, and the global community has facilitated this.

The fourth certainty is that individuals, private enterprises, and policy makers want to reduce waste. These goals may be either national or international. The final certainty is that the amount of waste being produced is increasing. Indeed, all of the five waste streams (wastes which come from the human body; wastes from everyday life which are common; and wastes which are hazardous, nuclear and industrial) are expanding, not shrinking, in production.

## 2. UNWANTED SOLID MATERIAL OF NO FURTHER USE

For the purpose of law, the defining feature of waste is that the producer of the waste no longer wants to possess it. For example, the Waste Framework Directive of the European Commission defines 'waste' as any substance or object which the holder discards or intends or is required to discard.[1] Similarly, the 1989 Basel Convention on the Control of Transboundary Movements of Hazardous Wastes and Their Disposal (the Basel Convention) defines waste as 'substances or objects which are disposed of or are intended to be disposed of or are required to be disposed of by

---

[1] Article 3(1). Directive 2008/98/EC (19 Nov. 2008) on Waste.

the provisions of national law'.[2] For the Organisation for Economic Co-operation and Development (OECD) wastes are materials that are intended for disposal.[3]

Outside of this relatively simple point, the definition of waste becomes more difficult as the issue of classification of whether a particular thing is waste or not is a much contested process. Lawyers, policy makers, scientists, non-governmental organisations and especially those who produce waste, all engage with the definitions of waste in great detail. The strong degree of interest in this topic is because whether something is waste, or not, will impact directly upon how it should be managed. Different management options directly relate to considerations of environmental suitability, political acceptability and financial cost. These considerations are most obvious with regard to whether the discarded item can be sent for recycling, reuse or recovery. If the discarded item is suitable for further use, and it is then linked into a management regime for its safe and effective further utilisation, it is not waste. Rather, it is a resource. However, before it is part of that management regime, it is waste. Thus, a discarded plastic bottle on the roadside is waste. When it is in a recycling bin going to an approved recycler, it is a resource. All of those involved or interested in waste have to grapple with such nuances. Politicians want to reclassify wastes as it makes them look like they are meeting their targets. Environmentalists want to reclassify wastes to ensure they are adequately dealt with, and business people want to classify wastes to ensure that they can make the best financial benefits from either recycling or disposing of them.[4]

One of the important considerations for the purpose of this book, derived from the above paragraph, is that waste must be solid, in the

---

2 Article 2. For the discussion of the Parties on this, see Decisions V/30 (1999); VII/33 (2004); VIII/27 (2005); and IX/27 (2008). Finally, note Goal 1. Annex to Decision BC-10/2. Strategic Framework for the Implementation of the Basel Convention for 2012–2021.

3 See Council Decision C(94) on Transfrontier Movements of Hazardous Wastes. Also, OECD (1998). *Guidance Document for Distinguishing Waste from Non-Waste*. (OECD, Paris). ENV/EPOC/WMP(98)1/REV1.

4 MacBride, S. (2012). *Recycling Reconsidered*. (MIT Press, Boston). 118–20; Ministry for the Environment (2012). *Consented Non-Levied Cleanfills and Landfills in New Zealand*. (MFE, Wellington). 7–11; UN Human Settlements Programme (2011). *Solid Waste Management in the World's Cities*. (UNSP, NYC). 6; Oelofse, S. (2008). 'Defining Waste'. *South African Journal of Science*. 104: 242–5; Wei, N. (2007). 'Re-Re-Re Defining Solid Waste'. *Pollution Engineering*. 39(8): 16–18; Cheyne, I. (2002). 'The Definition of Waste in EC Law'. *Journal of Environmental Law*. 14(1): 61–75; Porter, R. (2002). *The*

sense, it must have a weight. If a waste is light in weight and is not grounded by gravity, it becomes air pollution. This is a type of mobile waste that cannot be collected and held, and is dealt with by other legal regimes for air, ozone and climate change pollution. For the purposes of this book, waste is predominantly solid, but it may also be liquid, in either its original form, such as human waste, or in its final form, such as leachate. In terms of size considerations for solid types of waste there are no minimum size requirements. The small balls of hard micro-plastic (less than 5 millimetres in diameter) that end up as the residue from the larger plastic items that have disintegrated in the ocean (each litre-sized polyethylene terephthalate (PET) bottle can break down into over 10,000 granules of plastic) are probably the smallest types of waste, although the most commonly recognised small type of solid waste is the litter of the 4 trillion cigarette butts per year.[5]

## 3. THE COSTS OF DISPOSAL AND THEIR IMPLICATIONS

A second relative certainty about waste is that the costs of its disposal began cheaply, but did not stay there. This is because, in the beginning, there is often an excess of supply of land and/or there is little opposition to its being disposed of in land or water which is of little importance to most people. This was the pattern from the time of Antiquity until the 19th century. This only began to change in the 20th century, when communities began to object to waste being dumped in certain places, either on land, in rivers or the ocean, and thus a scarcity of places for dumping developed. At the same point, legislators began to impose restraints on what could be disposed of, and how. These restraints on where waste could be disposed of, and to what standards (if any) that this should be done, started to force up the economic costs of disposal. Nevertheless, with retrospect, the economic costs of disposal were very small. For example, in the 1930s, where nearly all waste was treated the same, the cost of disposal was 6 cents per cubic yard to send it to a landfill or to the incinerator. The going (commercial) price to dispose of

---

*Economics of Waste*. (Resources for the Future, Washington). 4; Wilkinson, D. (1999). 'Time to Discard the Concept of Waste'. *Environmental Law Review*. 1: 172–92.

[5] Novotny, T. (2014). 'Time to Kick Butts'. *New Scientist*. 28 June. 26; Baztan, J. (2014). 'The Hazards of Micro Plastic Pollution'. *Marine Pollution Bulletin*. 80: 1–2.

a 55 gallon drum of nuclear waste into the ocean in 1960 in the United States was 11 dollars.[6]

In the 21st century, the economic costs of disposal have sky-rocketed for both common and municipal waste and for that which is hazardous. The landfilling of standard municipal waste, averaged out per country, is about 7 dollars (US) per ton in the United States, about 70 dollars per ton in Europe, and 200 dollars per ton in Japan. In each area, the costs may be higher or lower, depending upon the ease with which the waste can be disposed of. This ease is directly related to the standards that have to be imposed before it can be dumped, and these standards are often related to considerations of proximity of human populations.[7]

If the wastes are hazardous, the economic costs become even more expensive. For example, in the developed world, the disposal of one ton of old pesticides is about 5,000 USD, whilst disposing mercury can cost 30 USD per kilogram. For high grade nuclear waste, the economic costs are near incomprehensible, as deep geological repositories cost tens of billions of dollars to complete, let alone monitor and ensure safety for the next million years.[8]

The implications of these rising economic costs of waste disposal, which were once a near invisible cost for consumers, municipalities and private or public corporations that generated waste, are that these costs are now becoming a dominant concern within budgets. Broad figures suggest that for most large scale manufacturers, the costs of waste disposal account for 3 percent of their total costs. For hospitals, it can be up to 10 percent of their budget. For cities in the developed world the

---

[6] Hamblin, J. (2009). *Poison in the Well: Radioactive Waste in the Oceans.* (Rutgers UP, New Jersey). 157; Rogers, H. (2005). *Gone Tomorrow: The Hidden Life of Garbage.* (Norton, NYC). 96, 80–81, 155; Rathje, W. (1992). *Rubbish: The Archaeology of Garbage.* (HarperCollins, NYC). 3–4, 86.

[7] UNEP (2012). *Global Environment Outlook 5.* (Progress Publishers, Malta). 173; UN Human Settlements Programme (2011). *Solid Waste Management in the World's Cities.* (UNSP, NYC). xxii–v; Mims, C. (2010). 'Landfills'. *Scientific American.* Sept. 52; Kinnaman, T. (2009). 'The Economics of Solid Waste Management'. *Waste Management.* 29(10): 2615–17; Bridgewater, A. (1975). 'A Systems Approach to the Economics of Waste'. *Resource Recovery and Conservation.* 1: 3–23.

[8] Joskow, P. (2009). 'The Economic Future of Nuclear Power'. *Daedalus.* 138(4): 45–59; Spoat, J. (2008). 'Nuclear Waste: Distant and Expensive Mirage'. *The Electricity Journal.* 21(7): 23–4; Porter, R. (2002). *The Economics of Waste.* (Resources for the Future, Washington). 54–5; Anon. (2002). 'Africa: Stockpiles of Obsolete Pesticides'. *Environmental Policy and the Law.* 32(6): 241–2; Pearce, F. (1998). 'A Wasted Chance'. *New Scientist.* May 30. 22.

economic costs of waste disposal may be up to 20 percent of their total budget. Globally, the public sector already spends more than 200 billion USD per year on waste disposal. This is expected to at least treble in the next 25 years.[9]

As the costs increase, individuals, entrepreneurs and/or policy makers continually try to find ways in which they can lower the economic costs imposed upon them, by finding cheaper methods of disposal. Assuming that the quantities cannot be reduced, the ways in which costs can be lowered are through illegally dumping the waste or finding the cheapest locations to dispose of the waste.

The illegal dumping of waste happens at the micro, national and international levels. At the micro level, those who avoid dealing with their waste correctly are known as litterers. Litter tends to be small amounts of waste that is out of place. In most parts of the developed world, the amount of litter is declining. This is because it is socially unpopular, heavily penalised and large amounts of money are spent by public authorities to keep streets clean. For example, in Britain the penalty for littering in a public place is up to £25,000. In addition, British local authorities also spend £350 million per year in keeping a small army of cleaners on the job.[10] At the national level, the illegal dumping of larger amounts of waste ranges from opportunistic individuals to organised crime. The scale of this problem is impossible to determine, as so many illegal waste dumps have not been located. As it stands,

---

[9] World Bank (2012). What a Waste: A Global Review of Solid Waste Management. (World Bank, Urban Development Series, No. 15, Washington). vii; Conrardy, J. (2010). 'Reducing Medical Waste'. *AORN Journal*. 91(6): 711; Franchetti, M. (2009). *Solid Waste Analysis and Minimisation*. (McGraw Hill, NYC). 3–4, 78–9; SPREP (2009). Pacific Regional Solid Waste Management Strategy: 2010–2015. (SPREP, Apia). 15. For some of the earlier attempts at economic forecasting in this area, see Chapter 21.16 of Agenda 21 (1992). UNCED. Doc A/CONF.151/4. Also reprinted in Johnson, S.P. (ed.). *The Earth Summit*. (Graham and Trotman, London). 21.

[10] Slavin, C. (2012). 'Linking Social Drivers of Marine Debris with Actual Marine Debris on Beaches'. *Marine Pollution Bulletin*. 64: 1580–88; Brown, T. (2010).'Picking up Litter: An Application of Theory-Based Communication'. *Journal of Sustainable Tourism*. 18(7): 879–900; Beck, R. (2007). *A Review of Litter Studies, Attitudes and Other Related Literature*. (Keep America Beautiful, Washington). 4; Jack, R. (2005). *Blowin' in the Wind: A Short History of Litter in the Twentieth Century*. (Arts and Humanities Council, London). 3–4; Huffman, K. (1995). 'Litter Reduction: A Review and Integration of the Literature'. *Environment and Behavior*. 27: 153; Seear, R. (1988). 'The Problems of Litter Control'. *The Journal of the Royal Society for the Promotion of Health*. 108: 46.

countries like Britain locate over 1,000 illegal dump sites per year, but the real number of illegal dump sites could be ten times this figure.[11]

At the international level, the illegal dumping of waste has been evident since the 1980s. This first became apparent in 1986 with the *Khian Sea*, a Liberian registered freighter, which was contracted to get rid of 15,000 tons of incinerator ash and other waste from Philadelphia, in the United States. The crew initially sailed to Haiti, where, with an agreement with the military government, they offloaded 3,700 tons of waste labelled as 'soil fertiliser', which was ash and pulverised glass, onto the beach. When it became apparent that the waste was not organic, the offloading stopped, and the ship escaped into the night with at least 10,000 tons of waste remaining on the vessel. The crew then tried to reach agreements with other developing countries – but all of them, now aware of the problem, declined them entry. When the ship was next seen, in Singapore, it had lost its cargo.[12] Despite this example being a large

---

[11] Gilbertson, K. (2012). 'Not a Load of Old Rubbish'. *The Safety & Health Practitioner*. 30(12): 44–6; Shinkuma, T. (2012). 'Effectiveness of Policy Against Illegal Disposal of Waste'. *Environmental Economics and Policy Studies*. 14: 123–45; Bisschop, L. (2012). 'Illegal Transports of E-Waste in Europe'. *Crime Law Society and Change*. 58: 221–49; Tompson, L. (2011). 'Profiling Illegal Waste Activity'. *European Journal of Criminal Policy and Resources*. 17: 179–201; Fukijura, J. (2011). 'Japan's Efforts Against the Illegal Dumping of Industrial Waste'. *Environmental Policy and Governance*. 21: 325–37; Matsumoto, S. (2011). 'The Effect of Community Characteristics on the Frequency of Illegal Dumping'. *Environmental Economics and Policy*. 13(3): 177; Ichinose, D. (2011). 'Waste Management Service and Illegal Dumping'. *Resource and Energy Economics*. 33: 79–93; Matsumoto, S. (2011). 'The Frequency of Illegal Dumping'. *Environmental Economics and Policy Studies*. 13: 177–93; White, R. (2011). 'Limitations in the Management of Hazardous Waste and Its Disposal'. *Journal of Environmental Protection*. 2(9): 1257–63; Liddick, D. (2010). 'The Traffic in Garbage and Hazardous Wastes'. *Trends in Organised Crime*. 13(2): 134–46; Ruffell, A. (2009). 'Illegal Waste Movement in Northern Ireland'. *Environmental Forensics*. 10: 208–13; Kim, G. (2008). 'Unit Pricing of Municipal Solid Waste and Illegal Dumping'. *Environmental Economics and Policy Studies*. 9(3): 167–80; Daele, S. (2007). 'Waste Management and Crime'. *Environmental Policy and the Law*. 37(1): 34–8; Block, A. (2002). 'Environmental Crime and Pollution'. *Social Justice*. 29(1): 61–72; Porter, R. (2002). *The Economics of Waste*. (Resources for the Future, Washington). 90, 92, 95–6; Gottinger, H. (1999). 'Crime, Control and Environmental Policy: Hazardous Wastes'. *Metroeconomica*. 50(1): 1–30; Fullerton, D. (1995). 'Garbage, Recycling and Illicit Burning of Dumping'. *Journal of Environmental Economics and Management*. 29: 78–91.

[12] Note, the waste in Haiti was repatriated in 1998 via a bilateral agreement with the United States. For general background, see Pellow, D. (2007). *Resisting*

catalyst for the formation of the Basel Convention, which tried to regulate the trade in waste, the problem of the illegal dumping of waste from one country in another continued throughout the 1990s and into the new century. The latest large scale incident of note occurred in 2006, when an unauthorised trade of 3,000 tons of petrochemical waste sludge, from France, made its way to the city of Abidjan on the Ivory Coast. At least seven people were poisoned to death by the waste, and some 40,000 people were reportedly made sick. These deaths were before France – even though it had not authorised the export – re-imported the waste.[13]

Assuming that those who wish to dispose of waste have some scruples and wish to follow the law and not dump illegally, the next step is to find where it is cheapest to legally dispose of waste. The cheapest manner to dispose of waste has always been at locations close to poorer communities which have less resistance to the waste and some may actually want it because of the economic opportunities it offers. These communities often have lower standards than richer areas. These lower standards are often because they are unaware of the risks, are not in a position to challenge the lower standards, or do not desire to set higher standards for waste management as it is not a pressing issue for them. These poorer communities are commonly dominated by racial and ethnic minorities, foreigners – and within this mix – often possess a strong gender element. Their unifying characteristic is that they are usually all well below the national levels for average income.[14]

---

*Global Toxics*. (MIT Press, Cambridge). 107–8, 119; Moyers, B. (1991). *Global Dumping Ground*. (Lutterworth, Cambridge). 16–22; MacKenzie, D. (1988). 'Britain Could Yet Take Italian Waste'. *New Scientist*. 8 Sept. 28; Anon. (1988). 'Toxic Waste Comes Home to Rest'. *New Scientist*. 9 June. 35.

[13] MacKenzie, D. (2006). 'Something Rotten in Africa'. *New Scientist*. 23 Sept. 8; Anon. (2006). 'French Take Ivorian Waste'. *New Scientist*. 11 Nov. 7. For the recording of this issue at Basel, see Statement by the President of the Conference of the Parties on the dumping of toxic wastes in Abidjan, Côte d'Ivoire, in August 2006. This is in Annex 3 of the 2006 (COP 8) Proceedings; for other examples of illegal dumping of an international flavour, see Edwards, R. (1995). 'Dirty Tricks in a Dirty Business'. *New Scientist*. 18 Feb. 12–13; Anon. (1992). 'Dumping Ground'. *New Scientist*. 20 June. 9; MacKenzie, D. (1992). 'Would-Be Waste Smugglers Face Execution'. *New Scientist*. 21 Nov. 8.

[14] Anon. (2014). 'The Colour of Pollution'. *Economist*. 24 May. 31; Roots, W. (2009). 'Environmental Movements, Waste and Waste Infrastructure'. *Environmental Politics*. 18(6): 817–34; Watson, M. (2005). 'Just Waste? Municipal Waste Management and the Politics of Environmental Justice'. *Local Environment*. 10(4): 411–26; Revathi, H. (2007). 'Race, Environmental Justice, and Interest Group Mobilizations'. *Western Journal of Black Studies*. 31(1): 50–57;

In legal theory, it was never meant to be this way. The ideal was, and is in all liberal, property-owning democracies, that all people who own property have equal rights. This means that all of those who create a nuisance, such as waste, are obliged to ensure that their nuisances do not impact upon others. This view was clearly articulated in 1705, when the English court, looking back on 200 years of jurisprudence declared, 'as every man is bound to look to his cattle, as to keep them out of his neighbour's ground; so he must keep in the filth of his house or office, that it may not flow in upon and damnify his neighbour'.[15]

Whilst the theory of nuisance law kept the wealthy landowners protected, the same could not be said for either public nuisances caused by solid waste or those impacting upon poorer individuals or communities who could not afford to keep the nuisances of others away from them – even if they did own the property. This was evident in the early part of the 19th century, as the courts increasingly began taking the view that citizens could not bring actions against those who were doing actions which were 'reasonable'. Here the law morphed from being near absolute, to being much closer to the law on negligence, in which those who were creating waste could rely on the defence that it was reasonable to do so, for the good of industry, society and the country in general. That is, nuisance law came to be morphed to not stand in the way of the waste created through the industrial revolution. Or, as Justice Willes stated in 1858, 'private convenience must yield to public necessity'.[16]

---

Brulle, R. (2006). 'Environmental Justice: Human Health and Environmental Inequalities'. *Annual Review of Public Health*. 27: 103–24; Luloff, A. (1998). 'NIMBY and the Hazardous and Toxic Waste Siting Dilemma: The Need for Concept Clarification'. *Society and Natural Resources*. 11: 81–9; Heiman, M. (1996). 'Race, Waste and Class: New Perspectives on Environmental Justice'. *Antipode*. 282: 111–21; Vittes, M. (1993). 'Factors Contributing to NIMBY Attitudes'. *Waste Management*. 13: 125–9; Furuseth, J. (1991). 'Communal Response to a Municipal Waste Incinerator'. *Landscape and Urban Planning*. 21: 163–71.

15 *Tenant v. Goldwin*, 92 Eng. Rep. 222, 2 Ld. Raym. 1089, at 1092 (K.B. 1705). For a discussion on the medieval period, see Abert, J. (2013). *An Environmental History of the Middle Ages*. (Routledge, London). 63–7.

16 *Hole v. Barlow*, 140 Eng. Rep. 1113, 4 C.B. (N.S.) 334, 335 (C.P. 1858). For commentary around this area, see Pontin, B. (2012). 'Nuisance Law and the Industrial Revolution: A Reinterpretation of Doctrine and Institutional Competence'. *Modern Law Review*. 75(6): 1010–36; Rosenthal, L. (2007). 'Economic Efficiency, Nuisance and Sewage'. *Journal of Legal Studies*. 36(1): 27–62; Paavola, J. (2002). 'Water Quality as Property: Industrial Water Pollution and Common Law in the Nineteenth Century United States'. *Environment and*

In practice, the private conveniences that had to yield were those of the poorer communities who had to learn to live with the new waste streams of the industrial revolution and the attempts to deal with them. In the context of this work, the first attempts to control wastes that local communities began to object to were landfill sites. These sites, despite the public necessity of getting waste out of the cities, were very unpopular from the beginning. The odour, noise, vermin and risks of fire (not counting the environmental risks that became known in the next century, such as with incineration) never made these attractive locations in any neighbourhood.[17] To this day, communities that are close to waste disposal sites have, on average, a 15 percent deflation on residential value – especially when it can be seen, heard or smelt.[18]

The economic costs, combined with the sheer undesirable nature of living next to waste disposal areas, have made them deeply unpopular in many parts of the developed world for over 100 years. For example, as the Chief Health Officer for Washington City warned in 1889, and every subsequent planning official in this area would concur thereafter, 'appropriate places for garbage are becoming scarcer year by year … Already the inhabitants in proximity to the public dumps are beginning to complain'.[19]

---

*History*. 8: 295–318; Brenner, J. (1974). 'Nuisance Law and the Industrial Revolution'. *Journal of Legal Studies*. 3: 403; McLaren, J. (1983). 'Nuisance Law and the Industrial Revolution: Some Lessons from Social History'. *Oxford Journal of Legal Studies*. 3: 155.

[17] Chemel, C. (2012). 'Odour-Impact Assessment Around a Landfill Site'. *Journal of Environmental Management*. 93: 85–94; Aristide, P. (2011). 'Are the Closed Landfills Recovered Habitats for Small Rodents?' *Urban Ecosystems*. 14(4): 699–710; Cook, S. (2008). 'Problem Bird Species on Landfill Sites'. *Environmental Management*. (2008) 41: 834–43; Sharp, T. (2007). 'On Rats, Refuse, and Recycling'. *Journal of Urban Health*. 84(5): 23–9.

[18] Gamper, S. (2011). 'Hazardous Waste Cleanup, Neighborhood Gentrification, and Environmental Justice'. *American Economic Review: Papers & Proceedings*. 101: 3, 620–24; Richard, C. (2010). 'Do Landfills Always Depress Nearby Property Values?' *The Journal of Real Estate Research*. 32(3): 321–39; Farber, S. (1998). 'Undesirable Facilities and Property Values: A Summary of Empirical Studies'. *Ecological Economics*. 24: 1–14; Furuseth, O. (1990). 'Impacts of a Sanitary Landfill: Spatial and Non-Spatial Effects on the Surrounding Community'. *Journal of Environmental Management*. 31: 269–77.

[19] As quoted in Rathje, W. (1992). *Rubbish: The Archaeology of Garbage*. (HarperCollins, NYC). 110. See also, Anon. (1915). 'Garbage Disposal Plant Not a Public Nuisance when Authorised by Law.' *Public Health Reports*. 30(48): 3507–11.

The difference between then and now is that the inhabitants who are presently complaining are much better informed, educated, conscious of their rights, and have developed their own identity: NIMBY (Not-in-my-back-yard).[20] In addition, and by comparison, they are much wealthier than both their ancestors and their temporal neighbours existing in poorer countries. This point is particularly important for, whilst waste disposal was largely a matter for national policy for most of the 20th century, by the end of that epoch, the options had become international, as large numbers of developed countries offered much cheaper disposal options for waste than the poorer communities in the developed world. These countries are often willing to sacrifice social and environmental considerations for the benefit of hard currency.[21]

These new global destination options are in full accord with the established pattern that, if it is legal, waste for disposal will always flow towards the locations where it is cheapest to dispose of. Moreover, the greater the income gap difference, the greater will be the emphasis on the movement of waste from the rich countries to the poorer countries.[22]

---

[20] Johnson, R. (2012). 'Don't Call Me NIMBY: Public Attitudes Toward Solid Waste Facilities'. *Environment and Behavior*. 44: 410; Martuzzi, M. (2010). 'Inequalities, Inequities, Environmental Justice in Waste Management and Health'. *European Journal of Public Health*. 20(1): 21–6; Chilvers, J. (2008). 'Power Relations: The Politics of Risk and Procedure in Nuclear Waste Governance'. *Environment and Planning A*. 40: 1881–900; Short, J. (2004). 'Some Principles for Siting Controversy'. *Journal of Risk Research*. 7(2): 135–52; Hoffman, S. (2001). 'Negotiating Eternity: Energy Policy, Environmental Justice, and the Politics of Nuclear Waste'. *Bulletin of Science Technology & Society*. 21: 456; Gawanda, K. (2001). 'Nuclear Waste Transport and Residential Property Values: Estimating the Effects of Perceived Risks'. *Journal of Environmental Economics and Management*. 42: 207–33; Blowers, A. (1999). 'Nuclear Waste and Landscapes of Risk'. *Landscape Research*. 24(3): 241–50; Hunter, S. (1995). 'Beyond NIMBY: Explaining Opposition to Waste Facilities'. *Policy Studies Journal*. 23(4): 601.

[21] Raddatz, L. (2012). 'Environmental Justice in Germany'. *The Professional Geographer*. 65(3): 495–511; Buekens, A. (2011). 'Waste Incineration, PVC, and Dioxins'. *Journal of Material Cycles and Waste Management*. 13: 190–97; Hsu, S. (2006). 'NIMBY Opposition and Solid Waste Incinerator Siting in Democratizing Taiwan'. *The Social Science Journal*. 43: 453–9.

[22] Brownwell, E. (2011). 'Negotiating the New Economic Order of Waste'. *Environmental History*. 16: 262–89; Kellenberg, D. (2010). 'Consumer Waste, Backhauling and Pollution Havens'. *Journal of Applied Economics*. 13(2): 283–304; Baggs, J. (2009). 'International Trade in Hazardous Waste'. *Review of International Economics*. 17(1): 1–16; Pellow, D (2007). *Resisting Global Toxics*. (MIT Press, Cambridge). 13; UNEP/Basel Secretariat (2007). *Waste Without*

Moreover, to many minds, this is not a bad thing, for as the former Vice President of the World Bank, Lawrence Summers explained in 1991:

> Shouldn't the World Bank be encouraging MORE migration of dirty industries to the lesser developed countries? I can think of three reasons … 1. A given amount of health impairing pollution should be done in the country with the lowest wages. I think the economic logic behind dumping a load of toxic waste in the lowest country is impeccable and we should face up to that. 2. I've always thought that underpopulated countries in Africa are vastly UNDER polluted … 3. The concern over an agent [pollutant] that causes a one in a million change in the odds of prostate cancer is obviously going to be much higher in a country where people survive to get prostate cancer … [Uppercase in original].[23]

Aside from the strong ethical concern of this type of advocacy, the fact is that Summers correctly identified a trend which does exist in the trade of waste between rich countries and poor countries. This trend is a subset of the larger problem of what are known as 'pollution havens', where dirty industry migrates from one country to another. Pollution havens, for all industries, occur when environmental standards coupled with a mix of other considerations, such as, amongst others, lower labour costs, tax incentives, trade liberalisation, the costs of transport, the price of land and compliance costs, combine to create a clear economic advantage to move one pollution intensive industry to another country. The creation of such havens is particularly strong in some industries, such as with the importation of waste, where the benefits of disposing of waste in one country are much clearer than in another. In this area, the evidence strongly suggests that waste imports increase for a country whose environmental regulations are much lower than their trading partners.

---

*Frontiers: Global Trends in the Generation and Transboundary Movements of Hazardous Wastes*. (Secretariat, Basel). 15–19; Beukering, P. (2006). 'Modelling and Analysis of International Recycling Between Developed and Developing Countries'. *Resources, Conservation and Recycling*. 46: 1–26; Orloff, K. (2003). 'An International Perspective on Hazardous Waste Practices'. *International Journal of Hygiene and Environmental Health*. 206: 291–302; Gbadegesinn, S. (2001). 'Environmental Racism, Toxic Waste and Eco-Catastrophe'. In Westra, L. (ed.). *Faces of Environmental Racism: Confronting Issues of Global Justice*. (Rowman, NYC). 190, 201; Clapp, J. (2001). *Toxic Exports*. (Cornell UP, NYC). 23; Atteh, D. (1993). 'The Political Economy of Environmental Degradation: The Dumping of Toxic Waste in Africa'. *International Studies*. 30: 277–301.

[23] Summers, L. Internal Memo. Reprinted in Pellow, D. (2007). *Resisting Global Toxics*. (MIT Press, Cambridge). 9.

These lower standards are nearly always related to the economic status of the country, and their willingness to tolerate problems which wealthier countries wish to divest themselves of. The end result has been the remarkable growth in the trade in certain types of waste between countries. For example, between the middle of the 1990s and the end of the first decade of the 21st century, the exports of waste plastics and ferrous waste between the United States and countries in Asia increased by nearly 1,000 percent and 500 percent, respectively.[24]

The exception to this trend where waste flows to the countries with lower economic costs is when particular waste streams are strictly controlled, such as with hazardous waste, in which cases, due to international prohibitions through the Basel Convention, the trend is actually reversed, with more hazardous waste being shipped to developed countries for (effective) disposal, than the other way round.[25]

## 4. THE INTERNATIONAL CONNECTION

The vast majority of the trade in waste throughout the 1970s (and thereafter) was between developed countries. Although there were often questions of the integrity and management of this trade, it was not seen as a large issue, as it was broadly assumed that developed countries could look after themselves and could make their own informed choices.[26]

---

[24] Kellenberg, D. (2012). 'Trading Wastes'. *Journal of Environmental Economics and Management*. 64(1): 68–87; Grether, J. (2012). 'Unravelling the Worldwide Pollution Haven Effect'. *Journal of International Trade and Economic Development*. 21(1): 131–62; Kellenberg, D. (2010). 'Consumer Waste, Backhauling and Pollution Havens'. *Journal of Applied Economics*. 13(2): 283–304; Cole, M. (2009). 'Institutionalised Pollution Havens'. *Ecological Economics*. 68(4): 1239–48; Levinson, A. (2008). 'Unmasking the Pollution Haven Effect'. *International Economic Review*. 49(1): 223–43.

[25] Baggs, J. (2009). 'International Trade in Hazardous Waste'. *Review of International Economics*. 17(1): 1–16.

[26] Anon. (1992). 'French Dither over PCBs'. *New Scientist*. 19 Sept. 5; Moyers, B. (1991). *Global Dumping Ground*. (Lutterworth, Cambridge). 68–9; Laurence, D. (1989). 'Transporting Waste in the European Community'. *Environment*. July. 12–17; Anon. (1989). 'PCBs Come Home to Roost'. *New Scientist*. 9 Sept. 5; Anon. (1989). 'Road Tankers Poison Cross-Border Relations in North America'. *New Scientist*. 3 June. 7; Anon. (1988). 'More Waste in Britain'. *New Scientist*. 23 June. 30; MacKenzie, D. (1988). 'Britain Could Yet Take Italian Waste'. *New Scientist*. 8 Sept. 28; Anon. (1988). 'Toxic Waste Comes Home to Rest'. *New Scientist*. 9 June. 35; World Commission on Environment and Development (1987). *Our Common Future*. (OUP, Oxford). 227; Lloyd, A.

However, in the 1980s, when the amount of waste in need of disposal expanded, the opportunity for international trade grew and the economic benefits of disposing of it in poorer countries accelerated. In so doing, the trade in many types of waste for final disposal changed orientation and started going towards the developing world. Exactly how much was being shipped was highly uncertain, although an indication of this as a growth market was evident in the rapid number of companies in the United States that exported waste, which grew from 12 in 1980 to 522 in 1989. At exactly the same time, as the quantity of waste was growing, questions were being raised in a number of developed countries over the types of wastes being shipped and the inability of many developing countries to adequately deal with them.[27]

The first manifestation of this concern was under the American President Carter when his administration discovered, and stopped, at least one attempt to ship hazardous waste abroad for final disposal in 1980. Carter then attempted to implement a policy whereby importing countries had to be fully aware of the nature of what waste they were importing, for which exporters would have to obtain a licence showing this consent. As such, the United States pledged itself to warn foreign governments of what waste it had banned (and why) and to support a worldwide notification scheme in this area. Carter also wanted to look into the possibility of creating a list of some wastes which could not be exported under any circumstances.[28]

Upon taking office, President Reagan rescinded this approach, seeking something less 'cumbersome' that he believed handicapped American exporters. He also tried to place more emphasis upon importing countries

---

(1983). 'Governments Fail to Report on Toxic Material'. *New Scientist.* 4 Aug. 329; O'Neill, P. (1982). 'Is Europe's Dioxin Waste Going East?'. *New Scientist.* 11 Nov. 342; Anon. (1981). 'Plugging the Toxic Waste Loophole'. *New Scientist.* 3 Dec. 653; Pearce, F. (1981). 'Government Promises Action on Toxic Imports'. *New Scientist.* 28 May. 540.

[27] Pellow, D. (2007). *Resisting Global Toxics.* (MIT Press, Cambridge). 107–8, 119; Clapp, J. (2001). *Toxic Exports.* (Cornell UP, NYC). 36–7; Moyers, B. (1991). *Global Dumping Ground.* (Lutterworth, Cambridge). 10–11, 16–22, 32–9, 75–80; MacKenzie, D. (1989). 'If You Can't Treat It, Ship It'. *New Scientist.* 1 Apr. 24; Anon. (1985). 'Waste Export Banned'. *New Scientist.* 14 Mar. 5.

[28] Joyce, C. (1981). 'Carter Signs Order to Control Hazardous Exports'. *New Scientist.* 22 Jan. 195; Joyce, C. (1980). 'US Tries to Curb Exports of Toxic Chemicals'. *New Scientist.* 19 June. 294; Anon. (1980). 'Third World's Toxic Wastes'. *New Scientist.* 21 Feb. 550; Anon. (1980). 'Third World Trades in Toxic Waste'. *New Scientist.* 7 Feb. 379.

making informed choices for themselves.[29] Accordingly, the Resource Conservation and Recovery Act was amended in 1984 and came to stipulate in section 262(52) that exports of hazardous waste were prohibited unless:

> (a) Notification … [of the proposed trade in waste] has been provided [to the importer],
> (b) The receiving country has consented to accept the hazardous waste,
> (c) A copy of the EPA [Environmental Protection Agency] Acknowledgment of Consent accompanies the… shipment…,
> (d) The hazardous waste shipment conforms to the terms of the receiving country's written consent…

This approach was then adopted by the Organisation for Economic Co-operation and Development (OECD) in 1984 for trade in hazardous waste between OECD countries.[30] Two years later in 1986, they extended these recommendations to cover trade in hazardous waste between OECD and non-OECD countries.[31] The members of the OECD agreed to:

> Prohibit movements of hazardous wastes to a final destination in a non-member country without the consent of that country and the prior notification to any transit countries of the proposed movements.
>
> Prohibit movements of hazardous wastes to a non-member country unless the wastes are directed to an adequate disposal facility in that country.[32]

The core of this system was that no trade was to occur without the consent of the appropriate authorities of the importing country, who were given full information of what was involved, and – importantly – unless the hazardous wastes were directed to adequate disposal facilities in the importing country. Two years later in 1988 (and updated in 1994), the OECD established a 'core list' of wastes that could be considered

---

[29] Joyce, C. (1983). 'US Steps up Trade in Toxic Wastes'. *New Scientist.* 11 Aug. 391; Anon. (1981). 'Reagan Rescinds Carter's Order on Hazardous Exports'. *New Scientist.* 5 Mar. 592; Anon. (1981). 'Reagan Opens the Door to a Dangerous Trade'. *New Scientist.* 17 Sept. 707.

[30] OECD, Decision-Recommendation C(83)180/FINAL on Transfrontier Movements of Hazardous Wastes (1 February 1984).

[31] Resolution C(85)100 on International Co-Operation Concerning Transfrontier Movements of Hazardous Waste (20 June 1985) and Decision-Recommendation C(86)64/FINAL on Exports of Hazardous Wastes from the OECD Area (5 June 1986). These are reprinted in 25 *International Legal Materials* (1986). 1010.

[32] Section I(iii) and (iv) of the 1986 Recommendation, as noted above.

hazardous, and therefore had to be controlled.[33] The European Community mirrored the OECD approach by only allowing trade in waste with developing countries if it was done with prior notification and full consent by the certain importing party.[34] This approach, whereby regulated trade in hazardous waste was permissible if consented to, was mirrored at the bilateral level between the United States and its neighbours to the north and south. Article 3 of the 1986 Free Trade Agreement with Canada stipulated:

> The Parties shall permit the export, import and transit of hazardous waste across their common border for treatment, storage, or disposal pursuant to the terms of their domestic laws, regulations and administrative practices, and the provisions of this Agreement.

A similar approach was adopted by the United States in its 1987 bilateral Agreement Regarding the Trans-boundary Shipments of Waste of Hazardous Materials and Hazardous Materials with Mexico.[35] This required the full and clear consent of the importing country, based upon clear information about what was being sent. This information had to set down any risks with the material being shipped, so that there could be no misunderstandings with the importers. Subsequent bilateral agreements were made between the United States and Malaysia in 1995, Costa Rica in 1997 and the Philippines in 2001. The difference with these subsequent agreements is that all of the same principles of full and informed consent are applied, but they are restricted to trade from the developing countries to the United States – not the other way round. Similar agreements exist with East Timor and Australia.

It was against this background of rapidly expanding trade in waste for final disposal between developed and developing countries that the United Nations Environment Program (UNEP) was asked in 1986 by a number of African countries to create a treaty on the export of hazardous waste. This proposal was supported by most countries, including the OECD, although there was considerable debate in the three years leading up to its conclusion over whether it would be a treaty of prohibition or regulation, and if regulation, what this would consist of and cover.[36] The

---

[33] Decision C(88)90/FINAL on Transfrontier Movements of Hazardous Wastes (27 May 1988).

[34] Directive 84/631/EC (20 Dec. 1984) on Transfrontier Shipment of Hazardous Waste.

[35] This is reprinted in 26 *International Legal Materials* (1987). 15.

[36] The UNEP starting point was the Governing Council Decision 14/30 (17 June 1987). The OECD Resolutions were C(89)1/FINAL on the Control of

result of three years of deliberation was the 1989 (Basel) Convention on the Control of Transboundary Movements of Hazardous Wastes and *Their Disposal* (my italics). The point about disposal was particularly important, as the Convention, as originally drafted, was primarily about the final disposal of wastes – not alternatives like recycling. This Convention has received positive affirmation from the international community at all of the Earth Conferences in 1992, 2002 and 2012.[37] This Convention also has a very high level of participation with over 180 Parties, with only a handful of countries choosing not to become a signatory to this treaty, of which the United States, despite originally signing it in 1990, is the most obvious.[38]

Although the drafters of the Basel Convention were mindful of the 'growing threat to human health and the environment posed by … the transboundary movement of hazardous wastes and other wastes',[39] they created a treaty that advanced regulation, not prohibition. This view came out clearly in the eventual title of the Convention, which explained it was for the 'control' – not the prohibition – of transboundary movements of hazardous wastes 'and their disposal'.[40] To achieve the goal of a well-controlled trade in hazardous waste between the Parties based upon high management standards,[41] three key points were established.

---

Transfrontier Movements of Hazardous Wastes (30 Jan. 1989), and Resolution C(89)112/FINAL on the Control of Transfrontier Movements of Hazardous Wastes (18–20 July 1989); Anon. (1989). 'Americans Balk at Toxic Waste Treaty'. *New Scientist.* 18 Mar. 27; MacKenzie, D. (1989). 'If You Can't Treat It, Ship It'. *New Scientist.* 1 Apr. 24; Anon. (1989). 'Waste Treaty Signed, but Hardly Agreed'. *New Scientist.* 1 Apr. 20; Anon. (1989). 'Rifts Threaten Waste Pact'. *New Scientist.* 4 Feb. 23; MacKenzie, D. (1988). 'Countries at Odds over Waste Treaty'. *New Scientist.* 19 Nov. 19; MacKenzie, D. (1988). 'Africa Wages War on Dumpers of Poisonous Waste'. *New Scientist.* 23 June. 30.

[37] For 2012, see paragraph 219 of RIO+20. A/CONF.216/XX. For 2002, see Section III.22 of the Plan of Implementation of the World Summit on Sustainable Development. A/CONF.199/20 (4 Sept. 2002). For the original Earth Summit in 1992, see Agenda 21 (1992). UNCED. Doc A/CONF.151/4. Chapter 20: 40–41.

[38] Clapp, J. (2001). *Toxic Exports.* (Cornell UP, NYC). 47, 55.

[39] Preamble. Basel Convention. Paragraph 2.

[40] The desire to prohibit, even if this was not the final goal, was noted in the preamble of the Convention. The preamble also noted that the parties were 'Convinced that hazardous wastes and other wastes should, as far as is compatible with environmentally sound and efficient management, be disposed of in the State where they were generated'. Paragraph 23. This desire was supplemented by the goal of minimising the generation of hazardous waste. See article 4(2)(d).

[41] Exports or imports to non-parties are not encouraged (article 4(5)) – although they are possible (through what are known as article 11 Agreements),

First, states have the right to say no to the disposal, or transit, of foreign wastes in or through their territory. States also have the right to say yes to the importation of foreign wastes.[42]

Second, for a trade in hazardous waste to be legitimate, it must follow strict procedures, or be repatriated. Specifically, each shipment must be clearly consented to.[43] If the terms of the consent are not followed, the waste must be taken back by the state of export. To help comply with the given consent, the waste must be packaged, labelled and identified correctly, carefully tracked, monitored, and covered by insurance, bond or other guarantee as may be required by the state of import or any state of transit. If there are mistakes of substance in the shipment (such as incorrect or wrongful labelling), the waste is deemed illegal and must be reimported by the exporting country. The adequate dealing with such illegal waste was seen as a priority for the Basel Convention.[44]

---

on the assumption that the same principles of environmentally sound management of the Convention still apply. This important exception is found in article 11. The Secretariat tries to monitor this, but this area, a large loophole through which bilateral agreements can bypass the stricter standards of the Basel Convention, is both hazy and vague. For the decisions on article 11 Agreements, see I/9 (1992); II/10 (1994); III/9 (1995); IV/1 (1998); V/20 and V/21 (1999); VI/17 and VI/18 (2002); VIII/28 (2006); and IX/28 (2008).

[42] For the rejection options, see articles 4(1)(a) and (b), 4(11) and 7. Note, states can also set higher standards than those established by the Basel Convention. For the acceptance, see article 4(1)(c).

[43] Specific consent (article 4(1)(c)); repatriation if consent not followed (article 8); correct labelling (article 4(7)(b)); financial coverage (articles 4(7)(a) and (c), 6(9) and 6(11)); re-importation, articles 4(3) and 9(1) to 9(4). Note, this rule does not apply if the fault is within the importing state. If it is not clear where the fault lies, the responsibility shall be shared between importing and exporting states. For the European rules on notification, consent, financial guarantee, and so on, see articles 4 to 10, 13 to 24 and annex I.A, B and C of Regulation 1013/2006/EC (14 June 2006) on the Shipments of Waste.

[44] To enable prosecutions, national laws must be created and implemented. This has been supported by model legislation, instruction manuals, guidance documents, and training activities to support capacity building in this area. See articles 4(4) and 9(5). For the decisions in this area on the illegal trade in waste, see Decisions I/2 and I/15 (1992); II/5 (1994); III/5 and III/6 (1995); IV/12 (1998); V/23 (1999); VI/16 (2002); VII/32 (2004); VII/34 (2005); VIII/24 and VIII/26 (2006); IX/23 (2008); and 10/13 (2011). For the consideration of illegal waste in the 1991 Bamako Convention on the Ban on the Import into Africa and the Control of Transboundary Movement and Management of Hazardous Wastes (the Bamako Convention) and the 1995 Waigani Convention to Ban the Importation into Forum Countries of Hazardous and Radioactive Wastes and to Control

Third, for this consent to be full and informed, detailed information must be exchanged between the exporter and importer, including the methods of disposal and the potential effects of the proposed movement on human health and the environment.[45] The importance of this is that, even if the waste is consented to, the Parties should take all appropriate measures to ensure the waste is disposed of in an environmentally sound manner. Thus, all practicable steps should be taken to ensure the protection of human health and the environment against the adverse effects which may result from such waste. If an exporting state has 'reason to believe' that the waste will not be disposed of in an environmentally sound manner, the export should not proceed. This obligation is checked off by the importing state having to confirm the existence of a contract and specifying the environmentally sound management of the waste will occur.[46] Article 4(8) adds to this. Specifically:

---

the Transboundary Movements and Management of Hazardous Wastes (the Waigani Convention), see article 9 in both of them.

[45] Article 4(2)(f). Note also article 7(c). The information required is in Annex V.A. This has been continually revised; Annex V.B, which deals with information on the movement of the information document. For the decisions on the documentation, with regard to notification, see I/21 (1992); II/16 (1994); III/16 (1995); and VIII/18 (2006). For the same broad principles in the Waigani and Bamako Conventions, see article 6 and Annex IV(A) – in both of them. Note, Krueger, J. (1998). 'Prior Informed Consent and the Basel Convention: The Hazards of What Isn't Known'. *The Journal of Environment and Development.* 7(2): 115–37.

[46] The contract requirement is in article 6(3). Ensuring safe disposal is in the preamble and articles 2 and 4(2)(b). Reason to believe otherwise (article 4(2)(e) and (g), as well as 4(8) and 4(10)). Note the rider of according to criteria to be decided on by the Parties. For the Bamako and Waigani Conventions, note article 4(j) for the same principle. The strongest example of this type of approach, and possible restriction on trade, is found in article 12 of the European Regulation 1013/2006/EC (14 June 2006) on the Shipments of Waste. Specifically, although shipment of the identified wastes for recovery, to countries with suitable standards, is appropriate, a number of objections to shipments of waste for recovery are possible. The core of these is that the shipment is not in accordance with the importer's national legislation relating to environmental protection, public order, public safety or health protection; non-compliance with earlier shipments, or if it does not make economic sense (as in the value of the recovered goods, having regard to economic and/or environmental considerations). Of particular note, according to article 49, no shipment is allowed if the environment or human health is not adequately protected. The adequacy of protection is found in Annex VIII, and covers the various technical guidelines adopted under the Basel Convention and/or the OECD.

> Each Party shall require that hazardous wastes or other wastes, to be exported, are managed in an environmentally sound manner in the State of import or elsewhere. Technical guidelines for the environmentally sound management of wastes subject to this Convention shall be decided by the Parties at their first meeting.[47]

Such technical guidelines on the safe disposal of multiple types of hazardous wastes, and disposal methods, although approved by the Parties to the Basel Convention are only recommended guidelines. They are not obligatory standards.[48] Rather, they, like the supplementing Regional Centres set up to facilitate best practices in this area and help capacity building, are of persuasive value only, in trying to help – not mandate – developing countries move towards sustainable practices in waste disposal. As of 2014, within Africa and West Asia, the regional centres were in Cairo, Dakar and Pretoria. Within Asia and the Pacific, they were in Beijing, Jakarta and Apia. In Central and Eastern Europe,

---

[47] Note also article 10(2)(e). For the decisions on the creation and use of the technical guidelines, see Decision I/19 (1992); II/13 (1994); III/13 (1995); VII/16 (2004); and IX/17 (2008).

[48] The guidelines on the final disposal overlap with Annex IV of the Convention which deals with 'disposal'. Disposal operations, which are given a 'D' code, are divided into those which do not lead to the possibility of resource recovery, recycling, reclamation, direct reuse or alternative uses, and those which do lead to such possibilities. The 15 disposal types range from deposit in the land, release into the ocean, biological treatment, through to incineration. These disposal options are accompanied by recommended technical guidelines. Similarly, guidelines exist for particular streams of hazardous waste. See, for example, Basel Secretariat (2012). *Updated General Technical Guidelines for the Environmentally Sound Management of Waste Consisting of, Containing or Contaminated with Persistent Organic Pollutants.* (Basel Secretariat, Basel); Basel Secretariat (2012). *Updated General Technical Guidelines for the Environmentally Sound Management of Waste Consisting of, Containing or Contaminated with Polychlorinated Biphenyls.* (Basel Secretariat, Basel); Basel Secretariat (2012). *Technical Guidelines for the Environmentally Sound Management of Wastes Consisting of Elemental Mercury and Wastes Containing or Contaminated with Mercury.* (Basel Secretariat, Basel); Basel Secretariat (2010). *Technical Guidelines for the Identification and Environmentally Sound Management of Wastes Resulting from Surface Treatment of Metals and Plastics.* (Basel Secretariat, Basel); Basel Secretariat (2003). *Technical Guidelines for the Environmentally Sound Management of Biomedical and Healthcare Wastes.* (Basel Secretariat, Basel). Note also the Guidance Document for the Environmentally Sound Management of Used and End of Life Computing Equipment. This was adopted through Decision BC/11/15 (2013). Partnership for Action on Computing Equipment.

they were in Moscow and Bratislava. For Latin America and the Caribbean, the locations were Buenos Aires, San Salvador and the Port of Spain (in the Caribbean).[49]

To secure compliance with the obligation of a safe and well-managed trade in hazardous waste, three supplementary tools have been developed. First, a fund has been created to cover accidents and emergencies involving the trade in hazardous waste under the Basel Convention.[50] Second, a Compliance Committee has been created;[51] and finally, the 1999 Protocol on Liability and Compensation for Damage Resulting from Transboundary Movements of Hazardous Wastes and Their Disposal was concluded.[52] This Protocol has the objective to provide for a comprehensive regime for liability and for adequate and prompt compensation for damage[53] for Parties acting under the Basel Convention 'due to

---

[49] See article 14 of the Convention. For the decisions, see VI/3 (1998). Establishment and Functioning of the Basel Convention Regional Centres for Training and Technology Transfer. Note, also, the earlier decisions around this area, notably, II/19 (1994); III/19 (1995); V/5 and V/6 (1999). Also, for the contemporary rules, see the following resolutions, VI/4 (1998); VII/9 (2004); VIII/4 (2006); IX/4 (2009); X/4; BC-11/12 (2013). Basel Convention Regional and Coordinating Centres; and BC-11/13. Process for Evaluating the Performance and Sustainability of the Regional and Coordinating Centres. For the locations, see VI/7 (2002); VI/8 (2002); VI/9 (2002); VII/10 (2004); and VIII/3 (2006).

[50] See article 14(2) of the Convention. For the decisions in this area for the Emergency Fund, see I/14 (1992); II/2 (1994); III/3 (1995); IV/20 (1998); V/18 – and the pivotal – V/32 (1999); VI/14 (1998); VII/29 (2004); and IX/22 (2008).

[51] See also Anon. (2003). 'Most Goals Achieved'. Environmental Policy and Law. 33(1): 16–17. See Decision VI/12 in particular. For its operation, see VI/13 (2002); VII/30 (2004); VII/31 (2004); VIII/31 (2006); VIII/32 (2006); IX/1 (2009); X/11 (2011) and X/12 (2011).

[52] The Protocol was foreshadowed in article 12 of the Convention. It took seven years to conclude. For the decisions on the negotiations, see I/5 (1992); II/1 (1994); III/2 (1995); IV/19 (2002); and BC-11/8 (2013). The Protocol is found in Decision V/30 (1999). Basel Protocol on Liability and Compensation for Damage Resulting from the Transboundary Movements of Hazardous Wastes and Their Disposal. Note also, Decision V/31 (1999). Relationship of the Protocol with the Basel Convention. For the reservations to the Protocol, see 'Report of the Fifth Meeting of the Conference of the Parties to the Basel Convention'. UNEP/CHW. 5/29. 10 Dec. 1999. Paragraphs 83–90. For the subsequent support for the Protocol, see Decisions VI/15 (2002); VII/28 (2005); VIII/25 (2006); and IX/24 (2008).

[53] According to article 2 of the Protocol, damage means loss of life or personal injury; loss of or damage to property other than property held by the person liable in accordance with the present Protocol; loss of income directly

an incident occurring during a transboundary movement of hazardous wastes and other wastes and their disposal, including illegal traffic, from the point where the wastes are loaded on the means of transport in an area under national jurisdiction of a state of export'.[54] Liability can be either strict (for the original person orchestrating the transfer) or fault based (once the waste has been deposited)[55] – or negated for contributory fault of the importers. The liability does not cover so-called aftercare incidents such as leaks or seepage at waste storage sites.[56]

Despite the efforts noted above, many of which only occurred towards the end of the 1990s, the initial years after the conclusion of the Basel Convention were full of the view that the trade in hazardous waste was more dangerous than ever, as trade expanded, morphed and became a high profile object of international concern that was seen as an unregulated trade where hazardous waste was being dumped upon misinformed or misled poor countries by those that were rich.[57] Or, more formally, as Agenda 21 recorded in 1992, there was an international concern that

---

deriving from an economic interest in any use of the environment, incurred as a result of impairment of the environment, taking into account savings and costs; and/or the costs of measures of reinstatement of the impaired environment, limited to the costs of measures actually taken or to be undertaken; and the costs of preventative measures, including any loss or damage caused by such measures, to the extent that the damage arises out of or results from hazardous properties of the wastes involved in the transboundary movement and disposal of hazardous wastes and other wastes subject to the Convention. The regime covers shipments under the Basel Convention, but not those organised under separate bilateral or regional arrangements, or complementary international agreements.

[54] Article 3.

[55] Articles 4 and 5.

[56] While the Protocol did not set out any financial limits for fault-based liability, it did set out minimum levels of compensation for strict liability. Although financial limits for strict liability were to be determined by national law, liability for notifiers or exporters for any one incident must be no less than 1 million SDR (Special Drawings Rights – about 1.3 million USD) for shipments up to 5 tons of hazardous waste, and then progressing up to 10 million SDR for shipments up to 10,000 tons. Beyond these amounts, an additional minimum of 1,000 SDR will be fixed for each additional ton of waste up to a maximum of 30 million SDR (41.4 million USD) for any one incident. See articles 12, 14 and 15, as well as Annex B.

[57] Edwards, R. (1995). 'Dirty Tricks in a Dirty Business'. *New Scientist.* 18 Feb. 12–13; Anon. (1992). 'Dumping Ground'. *New Scientist.* 20 June. 9; MacKenzie, D. (1993). 'Germany Defers Pesticide Ban'. *New Scientist.* 13 Mar. 6; MacKenzie, D. (1992). 'Would-Be Waste Smugglers Face Execution'. *New Scientist.* 21 Nov. 8; MacKenzie, D. (1992). 'Europe Continues "Poison My Neighbour" Exports'. *New Scientist.* 7 Nov. 8.

such trade was 'being carried out in contravention of existing national legislation and international instruments to the detriment of the environment and public health of all countries, particularly developing countries'.[58] Accordingly, Agenda 21 called upon the Parties to the Basel Convention, inter alia, 'to adopt a ban or prohibit, as appropriate, the export of hazardous wastes to countries that do not have the capacity to deal with those wastes in an environmentally sound way or that have banned the import of such wastes'.[59]

The concerns recorded in Agenda 21 quickly entered into the high-level debates between the Parties at the Basel Convention. When it was felt that this situation could not be adequately regulated, the majority of the Parties agreed, via resolution, to 'prohibit immediately all transboundary movements of hazardous wastes which are destined for final disposal from OECD to non-OECD States'.[60] This resolution was strongly objected to by many of the developed countries, on both philosophical and legal grounds. The latter argument was that such a fundamental change to the Basel Convention, making it one of prohibition, not regulation, of trade in hazardous waste for final disposal, could only be made by a formal amendment of the Convention (requiring a 75 percent majority) not a simple resolution (of a 50 percent majority).[61] Those seeking the prohibition followed suit, trying to formally amend the Basel Convention by inserting a new article (4A)[62] to prohibit the trade in

---

[58] Agenda 21 (1992). UNCED. Doc A/CONF.151/4. Chapter 20, paragraph 4.

[59] Agenda 21 (1992). UNCED. Doc A/CONF.151/4. Chapter 20, paragraphs 33 and 34.

[60] Decision II/12 (1994). Final Recovery Operations. Note also Decision II/14 (1994). Transboundary Movements of Hazardous Wastes Destined for Recovery Operations; Anon. (1994). 'Export of Hazardous Wastes'. *Environmental Policy and the Law*. 24(4): 147–8; MacKenzie, D. (1994). 'Europe Reluctant to Curb Toxic Waste Trade'. *New Scientist*. 2 Apr. 6.

[61] MacKenzie, D. (1992). 'Business as Usual for Traders in Toxic Waste'. *New Scientist*. 2 Dec. 9; MacKenzie, D. (1993). 'Europe's Green Channel for Toxic Waste'. *New Scientist*. 13 Mar. 13–14; Anon. (1995). 'Recycling of Hazardous Waste'. *Environmental Policy and the Law*. 25(4): 184–5; Edwards, R. (1995). 'Leaks Expose Plan to Sabotage Waste Treaty'. *New Scientist*. 18 Feb. 3–4.

[62] 'Each Party listed in Annex VII shall prohibit all transboundary movements of hazardous wastes which are destined for operations according to Annex IV A to States not listed in Annex VII. Each Party listed in Annex VII shall phase out by 31 December 1997, and prohibit as of that date, all transboundary movements of hazardous wastes under article 1(i)(a) of the Convention which are destined for operations according to Annex IV B to States not listed in Annex VII. Such transboundary movement shall not be prohibited unless the wastes are

hazardous waste for final disposal between developed and developing countries. However, although there were supporters of this view (receiving over 80 ratifications), they did not have the 75 percent majority required to formally amend the Convention.[63] Coupled with reservations about the proposed amendment due to the division of countries, the situation turned into one of extreme confusion. Although further work was set down to attempt to resolve the difficulties, all efforts in this area of actually implementing a ban on all hazardous waste for final disposal stalled, and then stopped.[64]

Work on breaking the deadlock began in 2009, when efforts were called for to create enabling conditions, through measures such as country-led initiatives that could help meet the objectives of the amendment. This drive culminated in work led by Indonesia and Switzerland in 2011, and a de facto, without prejudice entry into force of the Ban Amendment.[65] The rest of this wide ranging resolution called for, inter alia, the development of more guidelines and more working groups around the environmentally sound management of hazardous wastes, capacity building, and greater legal clarity on the definitions of, amongst

---

characterised as hazardous under the Convention.' Decision III/1 (1995). Amendment to the Basel Convention. See also Decision III/14 (1995). Transboundary Movements of Hazardous Wastes Destined for Recovery Operations.

[63] See article 17 of the Convention. If consensus is not possible, the amendment can be adopted by a three-fourths majority vote of the parties (article 17.3). Amendments adopted 'shall enter into force between Parties having accepted them … or acceptance by at least three-fourths of the Parties who accepted them or by at least two-thirds of the Parties to the protocol concerned who accepted them' (article 17.5). This issue and the words 'who accepted them' have been the subject of intense debate with regard to Decision III/1 over how best an amendment (and whether it should be by consensus, by entrenched majority, and if so – by whom) should be brought into force. See Decisions VIII/30 (2006) and IX/25 (2008), addressing the interpretation of paragraph 5 of article 17 of the Basel Convention.

[64] For attempts to keep the process going, see Decisions relating to Annex VII: IV/8 (1998); V/4 (1999); VI/34 (2002); and VII/24 (2004). For commentary, see Clapp, J. (2001). *Toxic Exports*. (Cornell UP, NYC). 77.

[65] Decision BC-10/3 (2011). Indonesian-Swiss Country-Led Initiative to Improve the Effectiveness of the Basel Convention. Specifically, the Parties agreed 'without prejudice to any other multilateral environmental agreement, that the meaning of paragraph 5 of article 17 of the Basel Convention should be interpreted to mean that the acceptance of three-fourths of those parties that were parties at the time of the adoption of the amendment is required for the entry into force of such amendment, noting that such an interpretation of paragraph 5 of article 17 does not compel any party to ratify the Ban Amendment'.

others, waste, non-waste, reuse, second-hand and used goods. This wide approach was followed up at the 2013 meeting, with the intention of resolving many of the difficulties in time for the 2015 meeting.[66]

The end result is that, at the international level, there is currently no general rule in the Basel Convention on the prohibition of the trade of hazardous waste for final disposal, to anywhere – except Antarctica.[67] Outside of the auspices of the Basel Convention, it has fallen to individual areas or regimes to develop their own prohibitions on the hazardous waste for final disposal. These prohibitions may be on the basis of either geography or specific pollutants which are governed by stand-alone international regimes.

In terms of restrictions on waste being placed for final disposal by geographical area, two approaches have been developed. The first involves restrictions on hazardous wastes being exported for final disposal. For example, Europe expressly prohibits the export of any type of hazardous waste for final disposal in other parts of the world.[68] In a complementary manner, two of the most economically vulnerable regions of the world, Africa and the Pacific, have treaties which explicitly prohibit the importation of any hazardous waste for disposal.[69]

In terms of restrictions by regime, three conventions are notable. Each of these has a strong relationship with the Basel Convention, in terms of cooperatively agreeing standards and setting guidelines for the particular wastes, whereby their work closely complements each other.[70]

---

[66] Decision BC-11/1 (2013). Follow-up to the Indonesian-Swiss Country-Led Initiative to Improve the Effectiveness of the Basel Convention.

[67] Article 4(6).

[68] See the European Regulation 1013/2006/EC (14 June 2006) on the Shipments of Waste. The prohibition is set down in articles 34 and 36. The covered wastes are those listed in their Annex V and are identified as hazardous explicitly, or declared so by the importing country. Annex V is the List A/Annex VIII from the Basel Convention. Anything on List A/Annex IX is prohibited for export. Anything from List B in the Basel Convention is also prohibited – if they are contaminated and can be identified as hazardous by any of the established procedures.

[69] The general obligation in Africa is found in article 2 of the Bamako Convention, whereby all parties promised to take all measures to prohibit the import of all hazardous wastes, for any reason, into Africa from non-contracting parties. Such import shall be deemed an illegal and criminal act. Article 4 of the Waigani Convention was exactly the same approach, but with the prohibition covering the South Pacific. Note also article 11 in both.

[70] For the relationship with the 1987 Montreal Protocol on Substances that Deplete the Ozone Layer and the Basel Convention, see Decisions II/15 (1994)

The first example is the 2001 Stockholm Convention on Persistent Organic Pollutants (the Stockholm Convention), which only allows a trade in this waste if it is 'destroyed or irreversibly transformed ... or otherwise disposed of in an environmentally sound manner'.[71] The Convention then adds that any trade must not be for any 'disposal operations that may lead to recovery, recycling, reclamation, direct reuse or alternative uses of persistent organic pollutants'.[72] This means that the selected pollutants can only be traded for the purposes of destruction, and in practice, the levels of destruction must be very high (a destruction efficiency of, at minimum, 99.99 percent). In practice, this means that this hazardous waste is only traded from developing to developed countries.[73]

Similar controls for the environmentally sound management of mercury-related waste were created in the 2013 Minamata Convention on Mercury, although given that mercury can be reused and/or recycled, the restrictions were not as tight. However, again, the expectation is that this trade is only from developing to developed countries in need of safe disposal or recycling options for this hazardous waste stream.[74] However, despite this goal, in practice, large amounts of mercury waste get into developing countries under the guise of material to be recycled, or as

---

and III/15 (1995). For the decisions on cooperation between the Basel Convention and the 1998 Rotterdam Convention on the Prior Informed Consent Procedure for Certain Hazardous Chemicals and Pesticides in International Trade and the 2001 Stockholm Convention on Persistent Organic Pollutants, see III/27 (1995); V/8 (1999) and V/9 (1999); IV/17 (1998); VI/29 (2002); VIII/8 (2006); X/2, X/10 and X/29 (2008).

[71] Articles 3(2)(a)(i), 3(2)(b)(i) and 6. Stockholm Convention. Note also SC-1/21 (2005). Technical Guidelines for the Environmentally Sound Management of Persistent Organic Pollutants Wastes.

[72] Article 6(1)(d)(iii).

[73] UNEP (2012). *Global Environment Outlook 5.* (Progress Publishers, Malta). 173; Pellow, D. (2007). *Resisting Global Toxics.* (MIT Press, Cambridge). 170; Karstensen, D. (2006). 'Environmentally Sound Destruction of Obsolete Pesticides in Developing Countries'. *Environmental Science and Policy.* 9: 577–86; Anon. (2004). 'Toxic Trade-Off'. *New Scientist.* 6 Nov. 7; Anon. (2002). 'Africa: Stockpiles of Obsolete Pesticides'. *Environmental Policy and the Law.* 32(6): 241–2; Bjorklund, E. (2000). 'Pressurised Liquid Extraction of Persistent Organic Pollutants'. *Trends in Analytical Chemistry.* 19(7): 434–50.

[74] Article 11. The Minamata Convention on Mercury.

within electronic consumer goods, and once it reaches the end of its useful life, it is not adequately collected and returned to the developed world.[75]

Finally, the 1987 Montreal Protocol on Substances that Deplete the Ozone Layer (the Montreal Protocol) controls the trade in ozone depleting chemicals for final destruction. Information in this area has always been collected, and the Parties have continually strived to develop the 'best technologies for … the destruction of controlled substances'.[76] Specifically, the destruction technologies for waste ozone depleting chemicals have been articulated since the early 1990s. These technologies have been continually reviewed and updated as new science has come to hand. The adopted technologies have been very high tech, including various forms of plasma, gas phase catalytic dehalogenation and/or super-heated steam reactor. Fundamentally, the original waste equipment carrying these chemicals must not be landfilled, whereby the chemicals are allowed to continue to slowly leak out to the detriment of the ozone layer. As with the above, the trade for final disposal (or recycling) is strongly of the orientation of being between developed countries, or from developing to developed countries, as only they possess the adequate technologies to achieve these goals.[77]

## 5. ZERO WASTE AND WASTE MINIMISATION

The fourth relative certainty about waste is that it is wasteful, and whilst a few people like recycling and encourage consumption because the

---

[75] Norom, I. (2011). 'Global Disposal Strategies for Waste Cathode Ray Tubes'. *Resources, Conservation and Recycling*. 55(3): 275–90; Mukherjee, A. (2004). 'Mercury in Waste in the European Union: Source, Disposal Methods and Risks'. *Resources, Conservation and Recycling*. 42(2): 155–70.

[76] Articles 4(a) and (b), 7 and 9(1)(a).

[77] For the appropriate Decisions in this area, see V/26 (1995); VII/35 (2000); XII/8 (2002); XIV/6 (2006); XV/9 (2008); XV/10 (2008); XVI/15 (2009); XVII/17 (2010); XVIII/9 (2011); and XXII/10 (2013). For the technologies, see Hodson, E. (2010). 'The Municipal Solid Waste Landfill as a Source of Ozone-Depleting Substances in the United States and United Kingdom'. *Atmospheric Chemistry and Physics*. 10(4): 1899–910; Heberlein, J. (2008). 'Thermal Plasma Waste Treatment'. *Journal of Physics*. 41(5): 78–89; Murphy, A. (2002). 'Plasma Destruction of Ozone Depleting Substances'. *Plasma Chemistry and Plasma Processing*. 22(3): 371–85. For the relevant decisions of the Montreal Protocol and the Basel Convention, mandating close cooperation, see Decisions V/24 (1995) and VII/31 (2000).

products can be recycled, no one likes waste and all of the environmental, economic and social costs that waste disposal creates. From these concerns, a movement has developed which seeks to stop the creation of the problem at the outset for, if there was no waste to dispose of, all of the problems associated with the disposal of waste would disappear. This is known as 'zero waste'.

The philosophy of zero waste is that a whole-system approach is required that aims to eliminate rather than 'manage' waste. Fundamentally, this system is based on the assumption that the impact of a good or service is much larger than the final result. That is, for the majority of products, over 90 percent of the materials and impacts a product causes never end up in the product that is for sale and are finally disposed of as waste. However, most waste strategies only see the end point, rather than the production process, as the focus. Zero waste aims to turn this around and incorporate both parts into deliberations about waste. In practice, this means examining and then aiming for with each manufactured product or process zero waste of resources (energy, materials and human), emissions (air, soil and water), activities (administration and production), toxins (in both processes and products) and product life – including transport, use and end of life. Thus, the goal of zero waste can be seen over the totality from the creation, operation and final disposal of the product.[78]

The goal of zero waste, especially in terms of manufacture, is of great importance to many commercial enterprises. Over 50 percent of the private sector in the developed world is actively engaged in trying to find ways to minimise their waste. They are active for reasons ranging from the offsetting of the costs of compliance, through to gaining a competitive advantage over their competitors. Most commonly, they wish to reduce waste because profit can be made in producing an improved good or service at less economic cost. This is very evident with a large number of consumer goods, all of which are of better quality, yet made with less materials and energy than their predecessors. For example, glass bottles, tin cans, aluminium cans, plastic bags, plastic bottles and yogurt pots,

---

[78] Zaman, A. (2014). 'Identification of Key Assessment Indicators of the Zero Waste Management Systems'. *Ecological Indicators*. 36(1): 154–72; Zaman, A. (2013). 'The Zero Waste Index'. *Journal of Cleaner Production*. 50: 123–32; Curran, T. (2012). 'A Zero Waste Vision'. *Journal of Hazardous Materials*. 207–8: 3–7; Lombardi, E. (2011). 'Zero Landfill is Not Zero Waste'. *BioCycle*. 52(7): 44–5; Staley, B. (2011). 'The Taste of Zero Waste'. *Waste Age*. Nov. 9–11; Kelleher, M. (2008). 'The Waste-Berg'. *Solid Waste & Recycling*. 13(1): 22–3; Anon. (2001). 'A Step Toward Zero Waste'. *Environment*. 43(4): 4–5.

all – per unit – produce less waste than earlier models because they all weigh less than half of what they did three decades ago. These decreases result in less material used, less energy needed to produce and transport the product, greater ease of recycling and less need for space in landfill.[79] In other instances, enterprise drives initiatives that will reduce waste for pure economic advantage. Thus, the enterprise that is at the forefront of making market-suitable plastic from a suitable resource, which easily degrades without a negative footprint, will make a fortune.[80]

Public authorities adopt a different strategy to zero waste than those who make the original products. This is because it is often the public authorities who have to deal with the product when it is at the end of its life and needs to be disposed of. These authorities have little influence on the design or production of the product. Accordingly, due to the limits of what they can actually achieve, the goals they commit themselves to tend to be much more about waste minimisation. Moreover, the goals that the public authorities adopt tend to become entwined with other considerations about particular types of waste destinations (with most wanting to avoid landfill) and/or complementary recycling targets. As such, their goals are not about zero waste overall, but rather, waste reduction or diversion in particular areas.[81] For example, ambitious targets of 50

---

[79] Nahman, A. (2010). 'Extended Producer Responsibility for Packaging Waste in South Africa: Current Approaches and Lessons Learned'. *Resources, Conservation and Recycling*. 54: 155–62; Pierce, J. (2008). 'Design Engineering: Green Payday'. *The Engineer*. Dec. 44–5; DeRamio, D. (2008). 'Green Giants Take on E-Waste'. *Communications News*. 45(8): 36; Tsoulfas, G. (2006). 'Environmental Principles Applicable to Supply Chains Design and Operation'. *Journal of Cleaner Production*. 14 (2006) 1593–602; Mohney, P. (2005). 'Design Goes Green'. *Machine Design*. 77(12): 64, 66, 68, 70–71; Cohen, E. (2004). 'Making Sense Out of Industrial Ecology'. *Journal of Cleaner Production*. 12: 1111–23; Ehrenfeld, J. (1997). 'Industrial Ecology: A Framework and Process Design'. *Journal of Cleaner Production*. 5(2): 87–95; Erkman, S. (1997). 'Industrial Ecology: An Historical View'. *Journal of Cleaner Production*. 5(2): 1–10.

[80] CBD (2012). *Impacts of Marine Debris on Biodiversity*. (CBD, Technical Series No. 67, Montreal). 4–5; Freinkel, S. (2011). *Plastic*. (Houghton, NYC). 209–12; Masui, A. (2011). 'Influence for Soil Environment by Continuing use of Biodegradable Plastic'. *Journal of the Polymer Environment*. 19: 622–7; Knight, H. (2010). 'New Plastic Recipe'. *New Scientist*. 3 July. 24. Anon. (2007). 'Boil the Bioplastic'. *New Scientist*. 28 Apr. 4; Anon. (2006). 'Plastic Bug's Genome Revealed'. *New Scientist*. 16 Sept. 27; Dean, F. (1995). 'Environmental Fads and Fallacies'. *Waste Management Resources* 13: 201.

[81] Zaman, A. (2014). 'Measuring Waste Management Performance Using Zero Waste Index'. *Journal of Cleaner Production*. 66: 213–31.

percent waste reduction in Canada (by the year 2000) and 55 percent in the Netherlands were rolled out in the early 1990s. In the new century, Britain adopted a number of Zero Waste Places in different regions ranging from small villages to entire cities. In New Zealand, almost half of the local authorities pledged themselves to eliminate all landfill waste by 2015. In Brisbane, Australia, the goal has been to reduce per-capita waste by 3 percent per year. Other cities, such as Los Angeles, established a goal of a 70 percent landfill diversion rate by 2013, followed by a 90 percent diversion rate by 2025. San Francisco is aiming to send zero waste to landfills by 2020, and the state of Florida has set a goal of recycling 75 percent of its waste by the same year.[82]

Such goals for minimisation have also been reflected at the regional level. For example, in Europe, all Member States have to propagate waste management plans and waste prevention programmes, with specific qualitative or quantitative benchmarks, taking into account the general goal 'to break the link between economic growth and the environmental impacts associated with the generation of waste'.[83]

At the international level, the goal of waste minimisation (not zero waste) found in many documents with some organisations, such as the OECD, recommending all of its members to 'foster waste prevention … and minimise quantities as well as potential risk of waste destined for disposal or recovery'.[84] More generalised recommendations also appeared in 1992 with Agenda 21, which called for all countries, when dealing with non-hazardous wastes to develop programmes 'to stabilize or reduce, if practicable, production of wastes destined for final disposal'.[85]

With regard to the goal of minimising the production of hazardous waste, the first international treaty to call for this was the 1989 Basel Convention on the Control of Transboundary Movements of Hazardous

---

[82] Krausz, R. (2013). 'Zero Waste: A Reality Check'. *Waste Management*. May. 8–9; Zaman, A. (2011). 'Urban Growth and the "Zero Waste City"'. *City, Culture and Society*. 2: 177–87; Phillips, T. (2011). 'A Critical Review of a Key Waste Strategy Initiative in England: Zero Waste Places – Projects 2008–2009'. *Resources, Conservation and Recycling*. (2011) 55: 335–43; Anon. (2008). 'Zero Waste, in Time'. *Environmental Management News*. Aug. 11.

[83] See articles 9, 28 and 29 of Directive 2008/98/EC (19 Nov. 2008) on Waste. These plans and programmes require public participation, evaluation and review. Note the value of Annex IV, on possible waste prevention measures.

[84] Recommendation 4. OECD (2007). *Guidance Manual on Environmentally Sound Management of Waste*. (OECD, Paris). 29.

[85] Agenda 21 (1992). UNCED. Doc A/CONF.151/4. See paragraphs 21.9 and 21.10(c).

Wastes and Their Disposal.[86] Three years later in 1992, Agenda 21 called for the 'stabilization of the quantity of hazardous waste generated' and 'establishment of long-term programmes and policies including targets where appropriate for reducing the amount of hazardous waste produced per unit of manufacture'.[87] This goal of minimisation was repeated in the regional waste conventions for both Africa and the South Pacific.[88] The goal to 'minimize and limit, where appropriate, the generation of radioactive wastes' was noted in Agenda 21,[89] as it was in the 1997 Joint Convention on the Safety of Spent Fuel Management and on the Safety of Radioactive Waste Management.[90] Finally, the need to minimise waste to avoid it being disposed of in the ocean was called for in the 1996 Protocol to the London Dumping Convention whereby the Parties agreed, according to their capabilities, 'to prevent, reduce and where practicable eliminate pollution caused by dumping or incineration at sea of wastes or other matter'.[91]

## 6. THE GROWTH IN PRODUCTION

The fifth and final certainty about waste is that despite the clear goals that exist locally, nationally and internationally to reduce the amounts of waste being generated, the exact opposite is occurring. The amount, and types, of waste that have to be dealt with are proliferating. Waste can broadly be divided into five types. These types are human; common; hazardous; nuclear; and dominant. Very simply, human waste is what comes from the body. Common waste is what people put in trash cans, and is not assumed to be, prima facie, dangerous. Hazardous waste is

---

[86] See paragraph 3 of the preamble, articles 4(2)(a), 4(13), 15(4) and (5). This goal has been repeated numerous times and extrapolated to specific waste streams. See IV/18 (1998); V/1 (1999); V/27 (1999); VII/2 (2005); VIII/23 (2006). Also the Cartagena Declaration on the Prevention, Minimization and Recovery of Hazardous Wastes and Other Wastes. This is in Annex IV, to the 10th Conference of the Parties, at the 2011 meeting in Cartagena; and Decision X/2 (2011): Strategic Framework for 2012–2021, and Objective 2.2 in particular.

[87] Agenda 21 (1992). UNCED. Doc A/CONF.151/4. Chapter 20, paragraphs 12(c) and (d). Note all of 20(6) and 20(11).

[88] For the Waigani Convention, see articles 4(4) and 10(2)(c). For the Bamako Convention, see article 4(3).

[89] Agenda 21 (1992). UNCED. Doc A/CONF.151/4. Chapter 22(4).

[90] Article 11(ii).

[91] Article 2. 1996 Protocol to the 1972 Convention on the Prevention of Marine Pollution by Dumping of Wastes and Other Matter.

dangerous and requires special handling. Nuclear waste is the most hazardous waste of all. Dominant wastes are those which produce the largest amounts of waste, by weight and size, of any sector in a society (such as dredging, mining or agriculture).

### A. Human Waste

Every single person on the planet is intimately concerned on a regular basis with the need to relieve themselves in a congenial place and fashion. This is necessary to emit slightly less than 100 grams of faeces and 1.5 litres of urine per day. This current amount of human waste per person of about 500 litres of urine and 50 kilograms of faeces per year is the same amount as it was thousands of years ago. However, the production of human waste is expected to grow rapidly in the coming decades.[92]

The reason this waste stream is increasing is twofold. First, between 2015 and 2050, the world population is expected to increase by 2.3 billion, passing from 7.0 billion to 9.3 billion. Second, the human numbers are growing in specific – urban – areas which necessitate particular management requirements for the problem of human waste.[93] Urban populations are projected to pass from 3.7 billion in 2012 to 6.3 billion in 2050. Asia is projected to see its urban population increase by 1.4 billion, Africa by 0.9 billion and Latin America and the Caribbean by 0.2 billion. The largest urban growth will occur in India, China and Nigeria. By 2050, India is projected to add 404 million additional urban dwellers. Nigeria will have an additional 212 million, whilst China will have to find space and infrastructure for a further 292 million.[94]

Within this mix, megacities (above 10 million people) will go from 23, in 2014, to 37 (of which 22 will be in Asia) in 2025. By 2050, six billion people will live in cities. Each one of these extra people will increase the total amount of excrement that has to be dealt with. Whilst some countries such as China are building infrastructure to deal with the

---

92 Black, M. (2008). *The Last Taboo: Opening the Door on the Global Sanitation Crisis*. (Earthscan, London). 3, 35–40.

93 For example, in Europe, with the implementation of their Urban Waste Water Treatment Directive, as all settlements above 2,000 people get brought within the obligation for suitable sewage requirements, the quantities of sewage doubled (from some 5 million tons of dry matter in 1992) as the number of households connected to the sewers increased. See Directive 91/271/EEC (21 May 1991) on Urban Waste Water Treatment.

94 United Nations (2014). *World Urbanisation Prospects*. (UN, NYC). 3–4.

increases in human waste faster than population growth, many developing countries are not. In these less well-planned environments, poverty, environmental degradation and lack of adequate infrastructure are leading to the formation of ever-larger slums, of which the inability to adequately deal with all forms of waste, and human waste in particular, is a defining feature.[95]

## B. Common Waste

Common, or municipal, waste is something that citizens produce, and this waste is dealt with collectively. It is typically assumed to be more of an annoyance than an environmental hazard. Although many hazardous wastes began as common wastes, until they are classified as hazardous, their environmental impact is considered to be of a lower risk. These are often nebulous distinctions, and are highly dependent on amounts generated, methods and locations of disposal.

The three certainties that are available with common waste is that the amount produced, per capita, is increasing in most countries; that this has been the pattern for the last 200 years; and that the type of common waste being produced is changing. The element of changing types has to do with changes in technology which directly impact upon consumer demand. For example, the 22 pounds of manure left by each horse per day in the bustling cities of the 19th century is no longer a problem as new types of transport have eclipsed this particular waste problem. However, in replacing one technology, different wastes have come to the fore. Good contemporary examples of this include packaging, plastic and electronic waste.[96]

---

[95] United Nations, Department of Economic and Social Affairs, Population Division (2012). *World Urbanization Prospects: The 2011 Revision.* (UN, NYC). 3–7; Buhaung, H. (2013). 'An Urbanization Bomb?'. *Global Environmental Change.* 23(1): 1–10; Pearce, F. (2013). 'Flushed with Success'. *New Scientist.* 16 Feb. 49; Chen, H. (2012). 'Utilization of Urban Sewage Sludge: Chinese Perspectives'. *Environmental Science and Pollution Research International.* 19(5): 1454–63; Katukiza, A. (2012). 'Sustainable Sanitation for Urban Slums'. *Biotechnology Advances.* 30(5): 964–78; Ling, O. (2007). 'Urbanization and Slum Formation'. *Journal of Urban Health.* 84(1): 27–34; Kwasi, B. (2005). 'Urbanisation Without Development: Environmental and Health Implications'. *Environment, Development and Sustainability.* 7(4): 465–500; Moore, M. (2003). 'Global Urbanization and Impact on Health'. *International Journal of Hygiene and Environmental Health.* 206(4): 269–78.

[96] Othman, S. (2013). 'Review on Life Cycle Assessment of Integrated Solid Waste Management in Some Asian Countries'. *Journal of Cleaner Production.*

In terms of historical growth of common waste, the trend has been obvious since the onset of the industrial revolution in the 19th century. This began as money became more available and a new class of people arose who desired, and were able to purchase, a variety of new products which were available, different and cheaper than previously. Accordingly, what was once restricted to the very wealthy would become the staple of most citizens. For example, the rubber tyre, invented by Charles Goodyear in 1839, would go on to average in the developed world (at the end of the 20th century) one tyre per person, per year. Yet for the first five decades that this product existed, only the wealthy could afford it. This type of growth was repeated in all sectors of the economy, leap-frogging each time there was a world war. This was particularly evident after 1945, when after the conflict ended the emphasis turned to consumer satisfaction and the creation of consumer demand with mass advertising, coupled with rising wages and credit facilities. Eight million motor vehicles (cars and commercial vehicles) rolled off the assembly lines in 1950. In 2014, it would be 45 million new vehicles per year, added to an existing fleet of over 1 billion. Supplementing the personal transport ideal, new and additional each year to the existing fleets, are nearly 1,000 aircraft and 1,000 ships.[97]

All such manifestation of this growth occurred, and continued to prosper as the industrial revolution morphed into the consumer revolution. This became apparent in 1955, when *Life* magazine heralded the advent of the throw-away society. Fashion, advertising, affordability, built-in obsolescence, convenience and an endless range of new things to purchase rapidly expanded what could be consumed, and subsequently,

---

41: 251–62; Anon. (2006). 'Less Fodder for Recycling Bins'. *New Scientist.* 16 Dec. 23; Rogers, H. (2005). *Gone Tomorrow: The Hidden Life of Garbage.* (Norton, NYC). 134–5; Strasser, S. (1999). *Waste and Want: A Social History of Trash.* (Holt, NYC). 13; Rathje, W. (1992). *Rubbish: The Archaeology of Garbage.* (HarperCollins, NYC). 103–5.

[97] Krausmann, F. (2009). 'Growth in Global Materials Use, GDP and Population During the 20th Century'. *Ecological Economics.* 68(10): 124–45; Olney, M. (1990). 'Demand for Consumer Durable Goods in 20th Century America'. *Explorations in Economic History.* 27(3): 234–56; Anon. (2012). 'Hard to Break Up'. *Economist.* 27 Oct. 26; Deshpande, P. (2012). 'Ship Recycling in India'. *Science of the Total Environment.* 438: 304–11; Xiang, J. (2011). 'Implementing Extended Producer Responsibility: Vehicle Remanufacturing in China'. *Journal of Cleaner Production.* 19: 680–6; Strasser, S. (1999). *Waste and Want: A Social History of Trash.* (Holt, NYC). 102–3; Hoy, S. (1996). *Chasing Dirt: The American Pursuit of Cleanliness.* (Oxford, OUP). 6–7, 174.

trashed.[98] This increase was so notable that in 1965 the American Congress noted:

> That the continuing technological progress and improvement in methods of manufacture, packaging, and marketing of consumer products has resulted in an ever-mounting increase, and in a change in the characteristics, of the mass of material discarded by the purchaser of such products; … the economic and population growth of our Nation, and the improvements in the standard of living enjoyed by our population, have required increased industrial production to meet our needs, and have made necessary the demolition of old buildings, the construction of new buildings, and the provision of highways and other avenues of transportation, which, together with related industrial, commercial, and agricultural operations, have resulted in a rising tide of scrap, discarded, and waste materials.

The associated waste from all of these sources was causing:

> [S]erious financial, management, intergovernmental, and technical problems in the disposal of solid wastes … solid wastes result in scenic blights, create serious hazards to the public health, including pollution of air and water resources, accident hazards, and increase in rodent and insect vectors of disease, have an adverse effect on land values, create public nuisances, otherwise interfere with community life and development.[99]

Despite these earlier concerns, the levels of production of common and/or consumer waste have continued to grow rapidly. Today, the cities of the world, in total, generate about 1.3 billion tons of solid waste per year. This volume is expected to increase to 2.2 billion tons by 2025. Waste generation rates are expected to more than double over the next 20 years in low income countries. Estimates suggest that the global per-capita average of municipal waste produced is about 438 kilograms per year (about 1.2 kilograms per capita, per day). Projections suggest that by 2025, some 4.3 billion urban residents will be generating about 1.42 kilograms per capita, per day of municipal waste (2.2 billion tons per year).[100]

---

[98] For the post-Second World War period, see the classic, Packard, V. (1959). *The Waste Makers.* (Pelican, London).

[99] Solid Waste Disposal Act. 79 Stat. 992 1965. Also, Public Law 89-272. 20 Oct. 1965. Section 202.

[100] World Bank (2012). *What a Waste: A Global Review of Solid Waste Management.* (World Bank, Urban Development Series, No. 15, Washington). viii, 2; Eurostat (2012). *Waste Statistics.* (EU, Brussels). 4; Karak, T. (2013). 'Municipal Solid Waste Generation, Composition, and Management: The World

Global figures on waste production hide the differences between the levels of waste, in terms of both total amounts and per-capita amounts, in the developed and developing world. In the developed (OECD) world, 572 million tons of common waste is produced per year. The per-capita values range from 1.1 to 3.7 kilograms per person, per day, with an average of 2.2 kilograms. These countries currently represent almost half (44 percent) of what is recorded for the solid (municipal) waste of the world. The per-capita figures for municipal waste have grown from 140 kilograms in 1940, 280 kilograms in 1960 and 440 kilograms in 1985. Today, this translates to a total of about 650 million tons of municipal waste per year in the developed world. The American average is just over 700 kilograms per person, whilst in Estonia, it is just over 300. In the American instance, the average amount of municipal waste disposed of by a citizen has nearly doubled (similar to many OECD countries) between 1960 and 2010, to a cumulative total of about 250 million tons of trash per year. In Europe (where the calculation includes industrial waste) inhabitants generated on average about 5.1 tons of waste each, of which 188 kilograms were hazardous waste.[101]

What is notable about the increased growth rates of waste production in the developed world is that the rate of increase is slowing in a few highly developed countries. Here, the current growth rate of municipal waste appears to be settling to about 1 percent increase per year. This growth is smaller than the growth rates in either gross domestic product (GDP) or expenditure on private consumption. Only a few countries have

---

Scenario'. *Critical Reviews in Environmental Science and Technology*. 42: 1509–630.

[101] World Bank (2012). *What a Waste: A Global Review of Solid Waste Management.* (World Bank, Urban Development Series, No. 15, Washington). 9; EPA (2011). *Municipal Solid Waste Generation, Recycling, and Disposal in the United States: Facts and Figures for 2010.* (EPA, Washington, EPA-530-F-11-005). 3-4; MacBride, S. (2012). *Recycling Reconsidered.* (MIT Press, Boston). 69; Franchetti, M. (2009). *Solid Waste Analysis and Minimisation*. (McGraw Hill, NYC). 35–7; Rogers, H. (2005). *Gone Tomorrow: The Hidden Life of Garbage*. (Norton, NYC). 2–3, 100–105; Pellow, D. (2007). *Resisting Global Toxics*. (MIT Press, Cambridge). 100–101; Porter, R. (2002). *The Economics of Waste.* (Resources for the Future, Washington). 24; IMO (1995). *Global Waste Survey*. (London, IMO). 8–9; Rathje, W. (1992). *Rubbish: The Archaeology of Garbage*. (HarperCollins, NYC). 102; Levenson, J. (1990). 'Wasting Away'. *Environment.* 32(2): 10–36; Pearce, D. (1976). 'The Economics of Waste Disposal in the UK'. *Resources Policy*. Dec. 276–85; Keen, B. (1977). 'Toxic Substances'. *New Scientist*. 15 Dec. 688; Packard, V. (1959). *The Waste Makers*. (Pelican, London). 21.

succeeded in actually reducing the growth rate in the quantity of solid waste generated for disposal. The few countries which are not following the trend and growing at a slower rate than other countries are high income, with very advanced reuse, recycling, an environmentally concerned citizenry and progressive solid waste disposal policies. Nevertheless, although not actually reducing the output, the evidence suggests that, above a certain level, further increases in GDP do not necessarily result in the increase of an equivalent amount of waste.[102]

This slowdown in the growth of common waste in some parts of the developed world is not the pattern in the developing world. Here, the production of municipal waste is rapidly growing and is showing no signs of plateauing. Currently, East Asia produces 21 percent (of which China represents 70 percent) of the world's municipal waste, and Latin America produces a further 12 percent. These figures are expected to increase rapidly, in accordance with growth rates of waste in the order of 30 percent in the last five years, in some regions. China has seen their per-capita waste production nearly double from 0.5 kilograms per person in 1980 to 0.98 kilograms in 2006. It is expected this will reach 1.8 kilograms per person by 2030. Cities like Beijing have quadrupled their waste production in the same period, going from a production of 1,040 million tons of waste to 4,134 million tons per year. This rate is expected to, at least, double by the year 2025. By 2030, it is expected that, per year, China will be producing twice as much municipal waste by volume and weight (but not the same per-capita amounts) as the United States. Similar projections exist for India. Delhi is expected to go from producing about 7,000 tons of municipal waste per day, to up to 25,000 tons per day by 2021. In Lagos, with a population of 21 million, over 10,000

---

[102] Wood, R. (2013). 'Going for Zero: A Comparative Critical Analysis of Zero Waste Events'. *Australasian Journal of Environmental Management.* 10(1): 46–55; Krausz, R. (2013). 'Zero Waste: A Reality Check'. *Waste Management.* May. 8–9; UNEP (2012). *Global Environment Outlook 5.* (Progress Publishers, Malta). 173; UNEP/Basel Secretariat (2007). *Waste Without Frontiers: Global Trends in the Generation and Transboundary Movements of Hazardous Wastes.* (Secretariat, Basel). 9–11; Editor. (2009). 'The Economics of Municipal Solid Waste Management'. *Waste Management.* 29: 2615–17; Mazzanti, M. (2008). 'Waste Generation, Waste Disposal and Policy Effectiveness: Evidence on Decoupling from the European Union'. *Resources, Conservation and Recycling.* 52: 1221–34; Johnstone, N. (2004). 'Generation of Household Solid Waste in OECD Countries: An Empirical Analysis Using Macroeconomic Data'. *Land Economics.* 80(4): 529–38; Porter, R. (2002). *The Economics of Waste.* (Resources for the Future, Washington). 1; Strasser, S. (1999). *Waste and Want: A Social History of Trash.* (Holt, NYC). 170–72.

tons of waste is already being generated per day. The end result of all of these figures is that in total, in terms of overall amounts (not per-capita) of common waste created, the developing world will produce a greater amount each year than the developed world by 2020, if not sooner.[103]

## C. The New Types of Common Waste

As alluded to above, not only is the amount of common waste increasing, so too is its type. This is most evident in three areas: packaging, plastic and electronics.

### (i) Packaging and plastic

Once a product is manufactured and ready to be sold, it must be distributed. To protect it from dirt and shocks, to make it easier to store and make it look appealing, a whole science has developed to design the most suitable wrappings. In the 20th century, this can be traced to the cardboard boxes that were introduced by Kellogg's in 1906, conveying its new product, cornflakes – safely wrapped in a waxed paper liner which sealed in its freshness. The first squeezable plastic bottle emerged in 1947. Aluminium foil packaging became common in the 1950s, and aluminium cans found their way to supermarket shelves in the 1960s. Disposable diapers, which arrived in the 1960s, have gone on (in the United States alone) to reach 18 billion units per year. These single use items, although occupying, on average, 1.5 percent by volume of all modern landfills, can last for 500 years before disintegration. Today, in the developed world, more than 30 percent of municipal waste is

---

[103] Anon. (2014). 'Clean It Up'. *Economist.* 22 Feb. 42; World Bank (2012). *What a Waste: A Global Review of Solid Waste Management.* (World Bank, Urban Development Series, No. 15, Washington). 1, 9; Zhang, L. (2012). 'Estimating Future Generation of Obsolete Household Appliances in China'. *Waste Management & Research.* 30(11): 1160–68; UN Human Settlements Programme (2011). *Solid Waste Management in the World's Cities.* (UNSP, NYC). xxvi, 11, 13; Zhou, L. (2011). 'Environmental Impact Assessment of Solid Waste Management in Beijing City'. *Waste Management.* 31: 793–9; Zhang, D. (2010). 'Municipal Solid Waste Management in China: Status, Problems and Challenges'. *Journal of Environmental Management.* 91(8): 1623–33; Giusti, L. (2009). 'A Review of Waste Management Practices and Their Impact on Human Health'. *Waste Management.* 29: 2227–39; Zhen, L. (2009). 'Municipal Solid Waste Management in Beijing City'. *Waste Management.* 29: 2596–9; Talyan, V. (2008). 'The State of Municipal Solid Waste Management in Delhi'. *Waste Management.* 28: 1276–87.

packaging, with the average consumer producing more than 150 kilograms of packaging waste per year.[104]

About 50 percent of disposed packaging is made of plastic. This material which emerged rapidly out of the 20th century is identified from the Greek verb *plassein*, which means to 'mould or shape'. Plastics are chemical combinations of additives and *polymers* (Greek for 'many parts'). They are substances made up of thousands of units called *monomers* (also Greek, for 'one part') linked into giant molecules. The original components of plastics are molecules, usually derived from oil and/or natural gas, that are bound to form polymers, which are then supplemented by various additives. The elements most commonly found in plastics are carbon, hydrogen, nitrogen, oxygen, chlorine, fluorine and bromine. It is estimated that around 4 percent of the annual world oil production is used as a feedstock for making plastics, and an additional 4 percent may be used in the generating process.

Plastics began when Alexander Parkes, in the middle of the 19th century, put his mind to finding an alternative to the very expensive ivory (then) necessary for billiard balls. Subsequent inventors such as Leo Baekeland refined the process, which by the turn of the 20th century was already proliferating in an ever increasing number of products and manifestations. Each new plastic product was light in weight, physically strong, resistant to other chemicals and the natural elements, and durable. They were also much cheaper to make than alternatives like ivory, rubber or shellac; and much more technologically malleable. This meant that the same worker who had turned out 350 hair combs per day in the 1930s could make more than 10,000 in the same period of time by changing to plastic and moulding machines. In the same period, Baekeland, who was featured in *Time* magazine as introducing 'the age of plastic', could boast his factory was churning out 8.8 million pounds of plastic per year. This

---

[104] MacBride, S. (2012). *Recycling Reconsidered.* (MIT Press, Boston). 175; OECD (2005). *Improving Recycling Markets.* (OECD, Paris). ENV/EPOC/WGWPR(2005)3/FINAL. 37–42; Franchetti, M. (2009). *Solid Waste Analysis and Minimisation.* (McGraw Hill, NYC). 26; Porter, R. (2002). *The Economics of Waste.* (Resources for the Future, Washington). 24; Rogers, H. (2005). *Gone Tomorrow: The Hidden Life of Garbage.* (Norton, NYC). 5; Morawski, C. (2003). 'Diaper Recycling'. *Solid Waste and Recycling.* 8(2): 6; Hoy, S. (1996). *Chasing Dirt: The American Pursuit of Cleanliness.* (Oxford, OUP). 174; Strasser, S. (1999). *Waste and Want: A Social History of Trash.* (Holt, NYC). 164–8; Rathje, W. (1992). *Rubbish: The Archaeology of Garbage.* (HarperCollins, NYC). 98, 155–65; Anon. (1991). 'Europe's Package Deal'. *New Scientist.* 14 Sept. 12; Marshall, J. (2007). 'Packaging Unwrapped'. *New Scientist.* 7 Apr. 37–40; Bell, J. (1990). 'Plastics: Waste Not, Want Not'. *New Scientist.* 1 Dec. 28.

market rapidly expanded during the Second World War as plastic became an essential item in military products, before going into overdrive, as mass consumerism settled in from the 1950s and the eight main branches of the plastic tree flowered into tens of thousands of different leaves, each satisfying a different demand.[105]

Perhaps the exemplar of the modern use of plastic is the plastic bag. Plastic bags are virtually without substance. They are evanescent puffs of polyethylene that are hygienic, odourless, waterproof, robust and convenient. Evolving out of the 1950s, this product, which already uses 75 percent less resin than 20 years ago, nearly made its competitor, the paper bag, extinct. This was due to them being three to four times cheaper to make than their heavier and bulkier opponent. As subsequent advocates of plastic bags would add, when viewed in terms of life-cycle analysis, plastic bags were better than the tree-fouling, chemical-pulping, intensive-bleaching, water-sucking industrial production processes required for paper bags. In addition, plastic bags possess the highest reuse rates of any disposable product – around 80 percent, going on to have a second life as lunch bags, bin liners, nappy bags, and so on. Due to such considerations, the production of lightweight plastic bags is monumental. Globally, between 500 billion and 1 trillion bags are carried home every year. That is, 150 bags a year for every person on Earth. Some people consume more than average. The annual per-capita consumption of plastic bags in the

---

[105] The current major types of plastics include high-density polyethylene (HDPE, typically for containers, toys, housewares, drums, milk and beer crates, shipping containers, industrial wrapping and film, gas pipes). This is the most common type of plastic, accounting for about one third of all plastics produced. HDPE is accompanied by low-density polyethylene (LDPE, for film, bags, toys, coatings, containers, pipes, cable insulation); polyethylene terephthalate (PET, such as for bottles, film, recording tapes, food packaging, synthetic insulation); polyester; polypropylene (PP, for film, battery cases, microwave containers, crates, car parts, electrical components, some food pots, such as for yogurt and margarine); polystyrene (PS, mainly for electrical appliances, thermal insulation, tape cassettes, cups, plates); poly-vinyl chloride (PVC, for window frames, pipes, flooring, wallpaper, bottles, cling film, toys, guttering, cable insulation, credit cards, medical products); expanded polystyrene (EPS, shock resistant packaging, cups and plates, thermal insulation, building components); and poly-tetrafluorethylene (PTFE, cable insulation, heat resistant coatings, electrical components, corrosion resistant fittings and seals). Freinkel, S. (2011). *Plastic*. (Houghton, NYC). 80–84; Basel Convention (2002). *Technical Guidelines for the Identification and Environmentally Sound Management of Plastic Wastes and for their Disposal*. (Basel Secretariat, Basel). 8–9; Rogers, H. (2005). *Gone Tomorrow: The Hidden Life of Garbage*. (Norton, NYC). 5; OECD (2005). *Improving Recycling Markets*. (OECD, Paris). ENV/EPOC/WGWPR(2005)3/FINAL. 43–7.

United States and Britain is about 300, whereas for other Europeans, the figure is closer to 200.[106]

Such consumption of plastic bags is only a subset of the larger pattern of plastic in modern society. Currently, the average American consumes about 30 pounds of plastic products per year. Conversely, average per-capita consumption in the developing world is about 10 pounds per year, but this is rising rapidly. The global production of plastic for both developed and developing countries exceeded 300 million tons in 2010. By 2050, projections suggest around 2 trillion pounds of plastic will be needed each year to satisfy global demand.[107]

The environmental costs of all plastic waste upon both the terrestrial and oceanic environments are put at 75 billion USD per year. These are the estimated financial impacts from issues such as the air pollution caused by their incineration, their costs in landfill, and their pollution impacts in the marine environment. Within landfill, plastic waste has gone from about 0.5 percent (by weight) of waste in 1960, to 12 percent (by weight, and some 30 percent by volume) around the end of the 20th century. Within this global average, there are large disparities. Plastic is only 2 percent (by weight) of the waste in Bamako, in Mali, whilst it is 17 percent in Rotterdam, in the Netherlands. India and China are rapidly moving towards levels of plastic waste which are commensurate with western countries.[108]

The growth rates of plastic waste in the ocean are greater than those on land. Plastic has gone from being 2.5 percent of all litter in the ocean in

---

[106] Freinkel, S. (2011). *Plastic*. (Houghton, NYC). 145; Moore, C. (2012). *Plastic Ocean*. (Avery, NYC). 32–41.

[107] Thanh, N. (2011). 'Assessment of Plastic Waste Generation and its Potential Recycling'. *Environmental Monitoring and Assessment*. 175: 23–35; Freinkel, S. (2011). *Plastic*. (Houghton, NYC). 1–7, 14, 68–9.

[108] UNEP (2014). *Valuing Plastic*. (UNEP, Nairobi). 2–4; MacBride, S. (2012). *Recycling Reconsidered*. (MIT Press, Boston). 181; Unnisa, S. (2012). 'Plastic Waste Management Strategies for Indian Cities'. *Journal of Chemical, Biological and Physical Sciences*. 2(1): 514–18; Thompson, R. (2009). 'Plastics, the Environment and Human Health: Current Consensus and Future Trends'. *Philosophical Transactions of Royal Society B*. 364: 2153–66; Pellow, D. (2007). *Resisting Global Toxics*. (MIT Press, Cambridge). 100; Rogers, H. (2005). *Gone Tomorrow: The Hidden Life of Garbage*. (Norton, NYC). 167; UN Human Settlements Programme (2011). *Solid Waste Management in the World's Cities*. (UNSP, NYC). 12; Basel Convention (2002). *Technical Guidelines for the Identification and Environmentally Sound Management of Plastic Wastes and for their Disposal*. (Basel Secretariat, Basel). 8–9.

1950 to over 80 percent of the debris in the ocean in the 21st century. Depending on what is counted (the options range from complete pieces of litter through to the broken-down pieces of micro-plastics), estimates range from 13,000 to 3.5 million pieces per square kilometre of ocean. Although some areas, such as the Antarctic, are relatively free of plastic, other oceans are not. In the Baltic Sea the density of litter is between 700 and 1,200 pieces per 100 metres of coastline, which is similar to the level found on the beaches of the northern North Sea. Some of the English coasts on the north-east Atlantic record levels of over 3,200 items per kilometre. Much of this litter is transported to the coasts of faraway places. Upwards of 60 percent of the plastic litter impacting upon New Zealand's coastal environment is not from New Zealand. Before such plastic is washed up, sinks or breaks up, it can accumulate on the high seas. The North Pacific Subtropical Convergence Zone, about the size of Texas and connected with 3 million pounds of debris, is believed to be the largest. By 2014, at least five other plastic gyres had been discovered, all centred around the thirtieth parallels north and south.[109]

This plastic waste in the ocean ranges from cigarette butts (made of cellulose acetate – not cotton), which take between 2 months and 15 years to degrade, through to abandoned synthetic fishing nets ranging up to 10,000 metres in length that can go on catching everything in their path for decades. This longevity is because, instead of biodegrading, most plastics photo-degrade, meaning they are broken apart by the ultraviolet radiation in sunlight. If very slowly destroyed by fungi, or faster by sunlight, the process is not quick, with many plastics having the ability to exist for hundreds of years. If submerged in the ocean (which becomes the fate for about 70 percent of all plastic types) many polymer molecules are virtually immortal. Even if they start to degrade, further problems can eventuate as they fragment into small particles. Between all of the forms that this plastic in the ocean may take, at least 267 species have been found to be detrimentally impacted upon by either choking

---

[109] Cozar, F. (2014). 'Plastic Debris in the Open Ocean'. *Proceedings of the National Academy of Science*. 111(28): 10029–40; Carson, H. (2013), 'The Plastic Associated Micro-Organisms of the North Pacific Gyre'. *Marine Pollution Bulletin*. 75(1); 143–53; Moore, C. (2012). *Plastic Ocean*. (Avery, NYC). 70–71; Mordecai, G. (2011). 'Litter in Submarine Canyons'. *Deep-Sea Research II*. 58: 2489–96; Barnes, D. (2010). 'Macroplastics at Sea Around Antarctica'. *Marine Environmental Research*. 70(2): 250–52; UNEP (2009). *Global Threat, Global Challenge: Review and Analysis of UNEP's Global Initiative on Marine Litter*. (UNEP, Nairobi); Ocean Conservancy (2009). *Marine Debris* (Ocean Conservancy, Washington). 3–4.

on it or by enhanced pollution levels, as the plastics often sequester pollutants, which are then ingested.[110]

**(ii) Electronics**

Products with electronics in them range from the portable communication devices we carry with us through to the domestic appliances in our homes. Such items, and the thousands between them, can be classified as Electrical Electronic Equipment[111] – or when at the end of their life, 'e-waste'. Such equipment is an essential part of modern life, and its production is rising exponentially. Computer sales are growing at more than 10 percent per year, with the world hitting the benchmark of 1 billion personal computers in 2002. Mobile phones have grown from less than 200 million users in 1997 to around 4 billion in 2012. This exponential growth is due not just to the satisfaction of needs, but also technological change, built-in redundancy and consumer lust. Mobile phones, like computers, have an average lifespan of about 2 years. Lifespans for such electronic items used to be more than double this less than a decade earlier. Assuming the e-waste is actually removed from the household and/or business (some 4 small items and 2.5 large items are believed to reside, unused, in most households in the developed world), then the resultant waste stream is particularly large. Globally, an estimated 20 to 50 million tons of electronic waste is being created each year. Europe produces about 9 million

---

[110] Scott, G. (2013). 'The Changing Nature of Oceanic Plastics Litter'. *International Journal of Environmental Studies*. 68(2). 191–5; Hammer, J. (2012). 'Plastics in the Marine Environment: The Dark Side of a Modern Gift'. *Review of Environmental Contamination and Toxicology*. 220: 1–44; Moore, C. (2012). *Plastic Ocean*. (Avery, NYC). 18–19, 86–8, 185–90; UNEP (2012). *Global Environment Outlook 5*. (Progress Publishers, Malta). 173; Freinkel, S. (2011). *Plastic*. (Houghton, NYC). 124–9, 132, 136–8; Shah, A. (2008). 'Biological Degradation of Plastics: A Comprehensive Review'. *Biotechnology Advances*. 26 (2008) 246–65; UNEP/FAO (2008). 'Abandoned, Lost or Otherwise Discarded Fishing Gear'. *UNEP Regional Seas Report*. 185: xiv–viii; Sul, J. (2007). 'Marine Debris Review for Latin America'. *Marine Pollution Bulletin*. 54: 1087–104; Edwards, R. (2005). 'Litter at Sea Means a Bellyful of Plastic'. *New Scientist*. 8 Jan. 11.

[111] Indicative list of EEE for the Europeans falls within the categories of Annex I, which covers a vast array of categories from toys to medical devices. More common examples include large household appliances, the list covers items such as refrigerators, washing machines, cookers, stoves. Computers and associated equipment of all varieties fit within the IT and telecommunications bracket, whilst consumer equipment covers everything from television sets to video recorders. See Annexes I and II of Directive 2012/19/EU (4 July 2012) on Waste Electrical and Electronic Equipment (WEEE).

tons per year. Japan produces around 4 million tons per year. In the United States, perhaps 400 million electronic items become obsolete each year, including 130 million cellphones and 47.4 million computers. The higher ends of the expansion in the growth of e-waste are in developing countries, which are projected to be producing twice as much e-waste as developed countries by 2020. Without imports, India currently produces 700,000 tons of e-waste per year, China 1.5 million tons and Nigeria close to 1 million tons per year.[112]

Accompanying the electrical waste is an endless array of batteries. The two basic types of household batteries are single use primary cells and rechargeable secondary cells. Among primary cells, the most used are the zinc-carbon and the alkaline-manganese batteries. Typically these batteries come in sizes AAA, AA, C, D and 9V. The quickly evolving rechargeable batteries, which have a much lower environmental footprint than their predecessor, are primarily nickel-cadmium, nickel-metal-hydride or variations of lithium. Silver oxide/button batteries are those found frequently in watches, toys and some medical devices. The exact amount of these batteries in use is a matter of speculation, although a global figure of 5 billion per year may not be far off. In Australia, an estimated 353 million portable batteries of all sorts are consumed each year. In the United States, about 3 billion alkaline batteries – alone – are purchased each year, or about eight per person each year.[113]

## D. Hazardous Waste

The most important definition by which all waste is divided is that which is hazardous, and that which is not. If a waste is deemed hazardous, it

---

[112] Anon. (2013). 'A Cadmium Lining'. *Economist*. 26 Jan. 52; UNEP (2012). *Global Environment Outlook 5*. (Progress Publishers, Malta). 173; Sthiannopkao, O. (2012). 'Handling E-Waste in Developed and Developing Countries'. *Science of the Total Environment*. 30: 1–15; Chi, X. (2011). 'Informal Electronic Waste Recycling'. *Waste Management*. 31: 731–42; Robinson, B. (2009). 'E-Waste: An Assessment of Global Production and Environmental Impacts'. *Science of the Total Environment*. 408(2): 172–87; Saphore, J. (2009). 'How Much E-Waste is There in the Basement'. *Journal of Environmental Management*. 90: 3322–31; Widmer, R. (2005). 'Global Perspectives on E-Waste'. *Environmental Impact Assessment Review*. 25: 436–58; Strasser, S. (1999). *Waste and Want: A Social History of Trash*. (Holt, NYC). 192, 209.

[113] Testsopang, S. (2008). 'Quantification and Characterization of Discarded Batteries'. *Resources, Conservation and Recycling*. 52(8): 1077–81; Bernardes, A. (2003). 'Collection and Recycling of Portable Batteries'. *Journal of Power Sources*. 124: 586–92.

must be identifiable as such, and it will be subject to completely different controls with regard to production, collection, transportation and disposal. In short, all of the requirements for management will be much more stringent than if it is non-hazardous. The degree of hazard will determine the level of management requirement. For example, the Basel Convention on *Hazardous* waste (my italics), and the multitude of obligations it creates, does not apply to non-hazardous materials.[114] The exact amount of hazardous waste that is produced each year is a matter of great uncertainty, as different types of waste, in different types of form, are counted. Nevertheless, the figure of 400 million tons per year is commonly used.[115]

By definition, the word 'hazard' refers to a situation that poses a level of risk. In this context, risk is the likelihood or probability that the properties of a waste will have a negative impact upon human or other ecological communities if the waste is not safely disposed of. Whilst this is easily stated, the difficulty is that human understanding of what types of waste can have a negative impact is continually evolving, as it has for the last 2,500 years. It is within this history that understandings of toxins, poisons and hazards emerged – and became overlapping in both law and policy. This emergence in law is most evident by the way in which the list of possible characteristics for a substance to be considered a hazard expanded. For example, in the 19th century, hazardous materials controlled under the law were those related to poison, and in the beginning, only one poison (arsenic) was controlled.[116] The list then expanded in the early part of the 20th century to cover materials that could cause fire, explode or corrode (such as in the 1927 Federal Caustic Poison Act of the United States).[117] By the 1960s, hazards were recognised due to their

---

[114] Vanguilder, C. (2012). *Hazardous Waste Management*. (Mercury, Boston). 49–50; Kabakci, B. (2005). 'A Rating Sytem for Determination of Hazardous Wastes'. *Journal of Hazardous Materials*. 126(1): 25–40; Dowling, M. (1985). 'Defining and Classifying Hazardous Wastes'. *Environment*. 27(3): 18–30.

[115] Eurostat (2012). *Waste Statistics*. (EU, Brussels). 4; Krueger, J. (1999). *International Trade and the Basel Convention*. (Earthscan, London). 12–15; Tolba, M. (1993). *The World Environment*. (Chapman, London). 265; World Commission on Environment and Development (1987). *Our Common Future*. (OUP, Oxford). 226; Milne, R. (1985). 'Hazardous Waste Threatens Environment'. *New Scientist*. 13 June. 3.

[116] The Act to Regulate the Sale of Arsenic (1851). 14 Victoria. Chapter 13, article 6.

[117] See section 2 of the 1927 Federal Caustic Poison Act. 44 Stat. 1406 (1925–27). For background, see Levy, J. (2011). *Poison: An Illustrated History*. (Lyons, Connecticut). 2, 18–19; Rodricks, J. (1996). *Calculated Risks: The*

being toxic, corrosive, ignitable or an irritant,[118] before these terms (and 'reactive') were clearly reiterated and defined in the 1976 Resource Conservation and Recovery Act.[119] To help work through these characteristics, the system in the United States was (and is) buttressed by five

---

*Toxicity and Human Health Risks of Chemicals in Our Environment.* (CUP, Cambridge). 15, 20–36, 48, 53, 62–3; Tarr, J. (1985). 'Historical Perspectives on Hazardous Wastes'. *Waste Management and Research.* 3(1): 95–102.

[118] See the Hazardous Substances Labeling Act. 74 Stat. 372 (1960). This Act defined the term 'hazardous substance' as any substance (or mixture) which was one of six criteria. These related to the substance being toxic (defined to mean any substance, other than a radioactive substance, which has the capacity to produce personal injury or illness to man through ingestion, inhalation or absorption through any body surface. Highly toxic was defined in terms of death rates for laboratory animals at given doses); corrosive (any substance which in contact with living tissue will cause destruction of tissue by chemical action; but shall not refer to action on inanimate surfaces); an irritant (any substance not corrosive within the meaning of the Act which on immediate, prolonged or repeated contact with normal living tissue would induce a local inflammatory reaction); a strong sensitiser (after taking into consideration the frequency and severity of reaction, would cause normal living tissue through an allergic or photodynamic process a hypersensitivity which becomes evident on reapplication of the same substance); was flammable (set at flash points above specific temperatures).

[119] The Resource Conservation and Recovery Act defines characteristics of hazardous waste. See Title 40 Code of Federal Regulations (CFR) Part 261, 'Identification and Listing of Hazardous Waste'. Ignitable wastes are liquids which have a flash point above 60°C (140°F) and compressed gases that possess a pressure that is flammable or could support combustion. Examples of these wastes are acetone, benzene, petroleum products, alcohols, naptha and used solvents such as xylene. Corrosive wastes are concentrated acids, bases or liquids that are capable of corroding metal containers, such as storage tanks, drums or barrels. Examples of such wastes are lead/acid batteries, etching solutions from photography/printing, ammonium solutions, hydroxide solutions, acid or alkaline cleaning solutions, rust removers, battery acid and caustic hot tank waste. Reactive wastes are unstable and can cause explosions, toxic fumes, gases or vapours when exposed to water, heat or increased pressure. Examples in this category include cyanide plating waste, waste containing bleaches, pressurised aerosol cans, metallic sodium and potassium. Toxic wastes are those which are harmful or fatal when ingested or absorbed through the skin. Toxicity is determined in the United States by the Toxicity Characteristic Leaching Procedure (TCLP). The results of this procedure are kept on what is known as the 'D-List'.

lists through which various types of waste having any of these characteristics within them are listed and classified as hazardous.[120]

Whilst such progress was to be commended, the difficulty became apparent that different countries were classifying hazardous wastes (and chemicals) differently. The international community began to address this problem as early as 1956 with their Globally Harmonised System of Classification of Chemicals. This system sought to establish a single harmonised system to classify chemicals, labels and safety data sheets. Repeatedly supported at the international level,[121] the Globally Harmonised System aims at ensuring that information on physical hazards and toxicity from chemicals is of the same language. The current regime covers 16 types of physical hazards,[122] 10 types of health hazards[123] and 2 types of environmental hazards.[124] Depending on the results of tests, the substance will be filtered into one of six divisions, which then results in differing symbols, signal words and hazard statements.[125]

---

[120] These are the D, F, K, P and U lists. The D-List, which currently contains 40 examples, is primarily based around established poisons (such as benzene), heavy metals (arsenic, cadmium, lead and mercury) as well as persistent organic pollutants (like 2,4-D, toxaphene and heptachlor). Accompanying the D-List is the F-List, which identifies 39 different types of hazardous waste from industrial and manufacturing processes (from petroleum refinery sludge through to spent halogenated solvents). The K-List covers over 100 hazardous wastes that come from industries that manufacture specific products, such as plastics or pesticides. The P- and U-Lists are unused commercial products, such as pesticides and pharmaceuticals that cannot be used or sold. See generally, Vanguilder, C. (2012). *Hazardous Waste Management*. (Mercury, Boston). 28–45.

[121] See paragraph 22(c) of the Plan of Implementation with the 2002 World Summit on Sustainable Development. Also, Agenda 21 (1992). UNCED. Doc A/CONF.151/4. Chapter 20, paragraphs 22(a), 32, 34(h) and (i).

[122] Explosives, flammable gases, aerosols, oxidising gases, gases under pressure, flammable liquids, flammable solids, self reactive substances and mixtures, pyrophoric liquids, pyrophoric solids, self-heating substances and mixtures, substances and mixtures that in contact with water emit flammable gases, oxidising liquids, oxidising solids, organic peroxides, corrosive to metals.

[123] Acute toxicity, skin corrosion/irritation, serious eye damage/irritation, respiratory or skin sensitisation, germ cell mutagenicity, carcinogenicity, reproductive toxicity, specific target organ toxicity – single and repeated exposures, and aspirational hazard.

[124] Hazardous to the aquatic environment, and hazardous to the ozone layer.

[125] See Chapter 2.1 of United Nations (2014). *Globally Harmonised System of Classification and Labelling of Chemicals*. (New York, UN). 43–51.

Although the international bodies dealing with hazardous waste keep a close watch on developments of chemical classification through the Globally Harmonised System,[126] the classification of hazards in general, from which hazardous wastes have been determined, has been much more influenced through the work of the OECD, which had identified 14 types of hazard.[127] Each hazard is given an H prefix and associated number. This H prefix and number is now common in many countries. For example, in Europe, the so-called 'H list' encompasses 16 types of characteristic of hazard, and like the approach of the United States, they too have provided lists through which various types of waste having any of the characteristics listed can be identified.[128] The most notable types of hazardous characteristics identified by the OECD are explosive (H1),[129] flammable liquids (H3),[130] flammable solids

---

[126] For the general obligation, see article 4(7)(b). For this cooperation with the United Nations Subcommittee of Experts on the Globally Harmonized System of Classification and Labelling of Chemicals, see Decisions VIII/19 (2006) and IX/20 (2008). For complementary work on cooperation with the World Customs Organization, see Decisions V/10 (1999); VIII/20 (2006); IX/19 (2008): and X/10 (2011).

[127] OECD Decision on Transfrontier Movements of Hazardous Wastes. 27 May 1988 – C(88)90/FINAL. Amended on 28 July 1994 – C(94)152/FINAL. The List of Hazardous Characteristics is in Table 5.

[128] For the characteristics, see article 3 and Annex III of Directive 2008/98/EC (19 Nov. 2008) on Waste. For the types, see Annex III of Directive 91/689/EEC (12 Dec. 1991) on Hazardous Waste. Annex I lists 40 different waste types, which are split into two parts. Annex IA deals with wastes which contain Annex III properties (ranging from biocides to resins, latex, plasticisers, glues/adhesives). Annex IB contains a list (ranging from untreated sewage to contaminated equipment) that contains any of the constituents listed in Annex II. Annex II, a list of 51 waste types that have the properties described in Annex III to render them hazardous, covers the commonly recognised heavy metals, persistent organic pollutants and various other problems, ranging from asbestos through to cyanides, acids and phenols.

[129] An explosive substance or waste is a solid or liquid substance or waste (or mixture of substances or wastes) which is in itself capable, by chemical reaction, of producing gas at such a temperature and pressure and at such a speed as to cause damage to the surroundings.

[130] The word 'flammable' has the same meaning as 'inflammable'. Flammable liquids are liquids, or mixtures of liquids, or liquids containing solids in solution or suspension (for example, paints, varnishes, lacquers, and so on, but not including substances or wastes otherwise classified on account of their dangerous characteristics) which give off a flammable vapour at temperatures of not more than 60.50 °C, closed-cup test, or not more than 65.60 °C, open-cup test (since the results of open-cup tests and of closed-cup tests are not strictly

(H4.1)[131] or liable to spontaneous combustion (H4.2).[132] Additionally there are hazardous wastes which emit inflammable gases when exposed to water (H4.3),[133] oxidise (H5.1),[134] cause acute poisoning (H6.1)[135] or can cause infection (H6.2).[136] Also hazardous are those which are corrosive (H8),[137] give off dangerous gases when exposed to air or water (H10), are toxic (H11),[138] ecotoxic (H12),[139] or which after disposal, may morph into another material, for example, leachate, which possesses any of the characteristics in any of the other H categories (H13).

The Parties to the Basel Convention (and the regional Bamako and Waigani Conventions[140]) use the H lists of the OECD, which are recorded in Annex III to the Basel Convention, to help define their

---

comparable and even individual results by the same test are often variable, regulations varying from the above figures to make allowance for such differences would be within the spirit of this definition).

[131] Solids, or waste solids, other than those classed as explosives, which under conditions encountered in transport are readily combustible, or may cause or contribute to fire through friction.

[132] Under normal conditions encountered in transport, or to heating up in contact with air, and being then liable to catch fire.

[133] Substances or wastes which, by interaction with water, are liable to become spontaneously flammable or to give off flammable gases in dangerous quantities.

[134] Substances or wastes which, while in themselves not necessarily combustible, may, generally by yielding oxygen, cause or contribute to the combustion of other materials. H5.2 covers accompanying organic peroxides – organic substances or wastes which contain the bivalent -0-0- structure are thermally unstable substances which may undergo exothermic self-accelerating decomposition.

[135] Liable either to cause death or serious injury or to harm human health if swallowed or inhaled or by skin contact.

[136] Substances or wastes containing viable microorganisms or their toxins which are known or suspected to cause disease in animals or humans.

[137] Substances or wastes which, by chemical action, will cause severe damage when in contact with living tissue, or, in case of leakage, will materially damage, or even destroy, other goods or the means of transport; they may also cause other hazards.

[138] These may be delayed or chronic, and turn on if they are inhaled or ingested or if they penetrate the skin, may involve delayed or chronic effects, including carcinogenicity.

[139] Ecotoxic substances or wastes are those which if released present or may present immediate or delayed adverse impacts to the environment by means of bioaccumulation and/or toxic effects upon biotic systems.

[140] See articles 2 and 3 of both the Bamako and Waigani Conventions.

wastes.[141] However, the H lists are not used in a pure form. Rather, they inform the decision making about Categories of Waste to be Controlled (Annex 1). Annex I contains a list of waste types which are recognised as being hazardous and in need of strict control.[142] Annex II comprises 'other wastes' that still require 'special attention'.[143] Annex I is informed by Annex III. From this combination of Annex I and Annex III, two lists were drawn up. These were known as List A and List B.[144] From such a basis, List A came to include those wastes for which trade between developed and developing countries required strict regulation, whilst List B includes wastes for which trade should be, prima facie, less regulated. In 1998, List A and List B were turned[145] into Annex VIII (covered by the Convention, and strictly regulated because hazardous)[146] and Annex

---

[141] See Annex III of the Basel Convention. As with every other national or regional body dealing with such lists, for Basel, each characteristic has to be verified by testing, of which the parties have directed work on key areas (such as infectious substances; toxicity; ecotoxicity) by bodies of the Convention. Most importantly, many of these characteristics have been subject to direct consideration to establish agreed levels of what are the *de minimis* limits in each setting. For the decisions on the hazardous characteristics, see V/24 (1999); VI/25 (2002); VII/17 (2004); VIII/21 (2006); and IX/18 (2008).

[142] Annex I pertains to waste streams that are to be controlled. Numbered as 'Y' categories, as of the time of writing, the list contains 45 types. These include wastes related to, inter alia, medicine, the pharmaceuticals, biocides, wood chemicals, explosives, resins, paints, glues, organic solvents, cyanides, industrial disposals, or materials containing polychlorinated biphenyls (PCBs), (polycyclohexylene terephthalates (PCTs) or polybrominated biphenyls (PBBs). Wastes which have particular constituents, such as heavy metals (including, inter alia, cadmium, mercury and lead), as well as cyanides, asbestos and various organic solvents are also in this bracket.

[143] Article 1(2).

[144] Hazardous waste may also be defined domestically. See article 1(1)(b).

[145] Decision IV/6 (1998). Outcome of the Technical Working Group on Lists of Wastes and the Applicable Procedure for Their Review or Adjustment. Also, Clapp, J. (2001). *Toxic Exports*. (Cornell UP, NYC). 93, 96–7. Note, although List A and List B were changed to Annex VIII and IX, the wastes on each list retain a fingerprint of their original classification, being an 'A' or 'B' waste, with an associated number. At the time of writing, 60 A waste types had been recognised, and 62 B waste types. Each recognised waste may have a subset of further waste streams within it.

[146] Annex VIII (List A) contains wastes which are characterised as hazardous under article 1(1)(a) of the Convention. This Annex covers many waste types, and prima facie applies, unless an exception exists in Annex IX. Annex VIII covers metal and metal-bearing wastes consisting of alloys, or containing the traditional heavy metals, including, inter alia, antimony, arsenic, beryllium,

IX (less regulated as not, prima facie, hazardous).[147] Needless to say, placement on either of these lists creates considerable debate as the obligations relating to a waste being on Annex VIII or IX are very different.[148]

---

cadmium, lead, mercury, selenium, tellurium and thallium. Various sludges, compounds or residues of zinc or copper, cyanide are also in Annex VIII. Notably, this part of Annex VIII also covers waste lead/acid batteries, whole or crushed, and scrap electronic assemblies (especially if it possesses Annex I wastes). Annex VIII also covers wastes containing principally inorganic constituents, which may contain metals and organic materials, such as, inter alia, glass from cathode-ray tubes, inorganic fluorine compounds, various gypsum, asbestos, and coal-fired power plant fly ash (in all instances, if they contain Annex I substances in concentrations sufficient to exhibit Annex III characteristics). Wastes containing principally organic constituents, which may contain metals and inorganic materials (such as that from the processing of petroleum, coke and bitumen, waste from the production of resins *unless* on List B, wastes containing biocides, wastes containing PCB, PCT, PCN or PBB, and asphalt waste. Medical waste, wastes from the creation of pharmaceutical or biocide products, those which contain cyanide, various acids and peroxides are also on Annex VIII.

[147] Annex IX (List B) contains wastes which are not covered by article 1(1)(a) of the Convention. That is, to be on Annex IX (List B), they have been determined (at the international level through the Basel processes) *not* to contain or are contaminated with Annex I materials to an extent causing them to exhibit an Annex III characteristic, making them hazardous (although they may still pose environmental or health risks when dealt with/disposed of). The parties have been monitoring this Annex IX quite closely. For the decisions on the national classification and control procedures for the import of wastes contained in Annex IX, and the additions to the list, see Decisions VII/20 (2004); VIII/22 (2006); IX/21 (2008); and BC-11/6 (2013). The broad parts of Annex IX (List B) are metal and metal-bearing wastes (especially in terms of precious metals – gold, silver, the platinum group, but not mercury – in pure, waste, residue or ash form), as well as, amongst others, scrap iron, steel, nickel, aluminium, zinc, tin, cobalt, chromium, titanium, magnesium and tungsten. Many clean, uncontaminated metal scrap, including alloys, in bulk finished form (sheet, plate, beams, rods, and so on) are also within this bracket (including, inter alia, scrap cadmium and lead – but excluding lead/acid batteries). Under this metal and metal-bearing category, electronic assemblies (not possessing Annex I or III considerations) are also listed, as are electrical and electronic assemblies (including printed circuit boards, electronic components and wires) destined for direct reuse, and not for recycling or final disposal. Finally within this metal bracket are listed waste end-of-life motor vehicles, containing neither liquids nor other hazardous components. The other parts of Annex IX (List B) cover wastes containing principally inorganic constituents, which may contain metals and organic materials, of which, amongst others, glass waste in non-dispersible form including cullet and other waste and scrap glass (except for glass from cathode-ray tubes and other

### (i) Persistent organic pollutants and medical waste

Two of the most well-recognised types of hazardous waste are persistent organic pollutants and medical waste. However, around both categories, there is considerable uncertainty about how much is being produced.

It is difficult to estimate exact quantities of pesticide wastes in existence. This is because many of the wastes are very old and documentation on where they are, let alone what quantities remain, is scarce. Moreover, different agencies often only see one part of the problem. For example, the Parties to the 2001 Stockholm Convention on Persistent Organic Pollutants only count obsolete pesticides that fall under their Convention – perhaps only 30 percent of the total of pesticides in existing waste dumps. Nevertheless, broad estimates suggest that there may be up to 600,000 tons of obsolete pesticides worldwide. Over 200,000 tons of these are understood to be in the former Soviet Union and its satellites. A further 290,000 tons are believed to be in Africa, with tens of thousands of tons in both Latin America and Asia. Although some of these waste pesticides are being removed, the problem continues to grow as pesticide use expands, as do the waste dumps of that which was unused. The problem with this type of calculation is that it not only ignores all of the other pesticides that are not within the ambit of the Stockholm Convention, but also, it ignores types of waste involving persistent organic pollutants that are not sitting in old containers, such as where they have been mixed into solvents such as oil, diluted with inert materials, found in old storage sites, old related equipment, packaging materials, soil, sediment and even human-related sludge. For all of these other types of wastes directly implicated with persistent organic pollutants, there is no estimate.[149]

---

activated glasses), ceramics, coal power plant fly ash if not listed on Annex VIII, and various industrial wastes at specific levels. Multiple types of plastic waste (provided they are not mixed with other wastes and are prepared to a specification) are covered, as is paper (and related products), textiles, rubber wastes (including waste pneumatic tyres, excluding those destined for Annex IVA operations), wastes from the agro-food industry, wastes from water-based paints, as well as resins and glues, and so on (not listed on Annex VIII) and even used single-use cameras, with batteries not included on Annex VIII.

[148] See article 1(1)(a). Also, Decision IV/9 (1998). Amendment and Adoption of Annexes to the Convention. For the decisions on the Review or Adjustment of Wastes Contained in Annexes VIII and IX, see Decisions VI/35 (2002); VII/19 and VII/21 (2004); and VIII/15 (2006). For background, see Krueger, J. (1999). *International Trade and the Basel Convention*. (Earthscan, London). 32–8.

[149] Agamuthu, P. (2013). 'Persistent Organic Pollutants in Solid Waste Management'. *Waste Management and Research*. 31: 967–71; Weber, R. (2013).

Medical waste is any solid waste that is generated in the research, diagnosis, treatment or immunisation of human beings or animals.[150] The average waste generation rates in the developed world are about 6 kilograms per bed, per day for general wastes. For infectious wastes, it is about 1 kilogram per bed, per day. Within this mix are wastes which are infectious, pathological, radioactive, biochemical, anatomical, contaminated sharps (i.e. needles, syringes, scalpels, and so on) and discarded medicines. On average, less than 10 percent of all medical waste is infectious and requires 'red bag' handling, due to risks such as being re-stuck with an infected needle (which carries risks of 30 percent, 1.8 percent and 0.3 percent for hepatitis B, hepatitis C and HIV, respectively). In some instances, due to specific initiatives, these percentages can change. For example, in 2001, a measles mass immunisation campaign in West Africa resulted in the generation of nearly 300 tons of injection waste. The cumulative total for both types of waste generated from medical establishments can be very high, such as the 4 million tons of medical waste generated in the United States each year. Whilst countries in the developed world are relatively well-equipped to deal with such waste, the same cannot be asserted in the developing part, where over 40 percent of surveyed countries do not have adequate disposal. The failures range from dumping the waste into the general stream of rubbish (where some medical waste can remain 'live' for decades) through to incineration of the medical waste being carried out inadequately. From

---

'The Need for an Integrated Approach to the Global Challenge of POPs Management'. *Environmental Science and Pollution Research.* 20(4): 1901–6; UNEP (2012). *Global Environment Outlook 5.* (Progress Publishers, Malta). 173. Galuszka, A. (2011). 'Pesticide Burial Grounds'. *Environment International.* 37: 1265–72; Dasgupta, S. (2010). 'Stockpiles of Obsolete Pesticides and Cleanup Priorities'. *Journal of Environmental Management.* 91: 824–30; Global Environment Facility (2010). *Reducing the Human and Environmental Risks of Obsolete Pesticides* (GEF, Washington). 4–5; Haylamichael, I. (2009). 'Disposal of Obsolete Pesticides'. *Environment International.* 35: 667–73; Pellow, D. (2007). *Resisting Global Toxics.* (MIT Press, Cambridge). 162–4; Dalvie, M. (2006). 'Disposal of Unwanted Pesticides'. *Science of the Total Environment.* 361: 8–17; Anon. (2004). 'Warning over Toxic Stockpile'. *New Scientist.* 18 Sept. 4; Anon. (2001). 'Poisonous Legacy'. *New Scientist.* 14 May. 7; Anon. (2002). 'Africa: Stockpiles of Obsolete Pesticides'. *Environmental Policy and the Law.* 32(6): 241–2; Clapp, J. (2001). *Toxic Exports.* (Cornell UP, NYC). 66.

150 Article 2 of the now lapsed Medical Waste Tracking Act. (1988) 42 USC. Paragraph 6992.

air pollution to the spread of disease, the risks of the inappropriate disposal of medical waste are very large.[151]

## E. Nuclear Waste

Unwanted nuclear material is the most dangerous of all types of hazardous waste. Unlike other types of waste, nuclear waste can be liquid or solid, as well as gaseous. For the purpose of definition, it is waste of which no further use is foreseen and it is controlled as radioactive waste by a regulatory body.[152]

Three principal classes exist by which nuclear waste may be divided. The first classification is that which is exempt from regulation. This type has such a low concentration of radionuclides that it can be excluded from regulatory control because the radiological hazards are considered negligible. Examples of this waste are often found in the tailings generated from the mining of uranium. Similarly, nuclear power stations and reprocessing plants commonly release small quantities of radioactive gases (such as krypton-85 and xenon-133) and trace amounts of iodine-131 to the atmosphere. These are chemically inert, have short half-lives (rates of decay) and their radioactivity is rapidly diminished by delaying their release. As such, their net effect is too small to warrant consideration in any life-cycle analysis.

---

[151] Mochungong, P. (2015). 'Assessing Health Risks from Sub-Standard Medical Waste Incineration'. *Human and Ecological Risk Assessment.* 21: 129–34; Shareefdeen, Z. (2012). 'Medical Waste Management and Control'. *Journal of Environmental Protection.* 12: 1265–8; World Health Organization (2011). *Health-Care Waste Management.* (WHO, Geneva, Fact Sheet No. 281). 3; Kahn, M. (2011). 'The Socio-Economic Impact of Improper Hospital Waste Management'. *Pakistan Journal of Medical Research.* 50(4): 214–34; Patwary, M. (2011). 'Scavenging and Recycling of Medical Waste'. *Journal of Environmental Management* 92: 2900–906; UN Human Settlements Programme (2011). *Solid Waste Management in the World's Cities.* (UNSP, NYC). 8–9; Cheng, Y. (2009). 'Medical Waste Production at Hospitals'. *Waste Management.* 29: 440–44; Huang, Y. (2006). 'Pyrolisis and Oxidation of Medical Wastes'. *Environmental Technology.* 27: 153–8; Amendola, A. (2007). 'Viable Newcastle Disease Vaccine Strains in a Pharmaceutical Dump'. *Emerging Infectious Diseases.* 13(2): 1907–10; Birchard, K. (2002). 'Out of Sight, Out of Mind. The Medical Waste Problem'. *The Lancet.* 359 (Jan.): 56; Awe, Y. (1996). 'Legislative and Scientific Aspects of Waste Disposal in Hospitals'. *Indoor and Built Environment.* 5: 6–21.

[152] Article 2(h) of the 1997 Joint Convention on the Safety of Spent Fuel Management and on the Safety of Radioactive Waste Management. Note also, IAEA (2013). *Nuclear Waste.* (IAEA, Vienna). 2.

The second classification covers low and intermediate-level waste. These types of waste hold enough radioactive material to justify safety standards for short or extended periods of time. The low-level waste can be that generated from hospitals and industry, such as paper, rags, tools, clothing and filters, or demolished material produced during rehabilitation or dismantling operations on nuclear industrial sites. This waste shares the common characteristic of having half-lives that are all 'short lived', being less than 30 years. This waste comprises some 90 percent of the volume but only 1 percent of the radioactivity of all radioactive waste. Intermediate-level nuclear waste is considered 'longer lived' as the radionuclides have half-lives greater than 30 years. Resins, chemical sludges and metal fuel cladding as well as contaminated materials from reactor decommissioning are commonly in this mix. This waste makes up some 7 percent of the total volume of all nuclear waste and has 4 percent of all the radioactivity held by the total of all radioactive waste.[153]

High-level nuclear waste arises from the 'burning' of uranium fuel in a nuclear reactor and the solidified liquid wastes from reprocessing such fuel to produce plutonium for nuclear weapons. This includes irradiated reactor fuel, liquid (or solidified) wastes from the first extraction cycle of reprocessing, and any other wastes of concentrations exceeding specified quantities of alpha emitters, beta/gamma emitters with half-lives greater than one year, and tritium and beta/gamma emitters with half-lives of one year or less. This high-level waste contains about 99 percent of the radionuclides and thus represents the largest radiological risk. This waste has both long-lived and short-lived components, with half-lives ranging from seconds to millennia. At the lower end of the spectrum, strontium-90 has a half-life of 29 years and caesium-137 has a half-life of 30 years. At the higher end, plutonium-239 has a half-life of 24,400 years. Technetium-99 has a half-life of 220,000 years. The half-life of neptunium-237 is 2 million years, whilst iodine-129 possesses a half-life of 17 million years. Whilst the half-lives required for the radioactivity of high-level waste to decay may be millions of years, it typically takes at least 10,000 years for the radioactivity of such waste to decay to the level which would have been generated by the original ore from which the nuclear fuel was produced, should this ore never have been mined.[154]

---

[153] Tavcar, P. (2007). 'Radiological Characterization of Low- and Intermediate-Level Radioactive Wastes'. *Journal of Radioanalytical and Nuclear Chemistry.* 273(3): 593–6.

[154] Vandenbosch, R. (2007). *Nuclear Waste Stalemate.* (Utah UP, Salt Lake). 20–21; Anon. (1997). 'High-Level Nuclear Wastes'. *Environment.* 39(3): 8.

The total amount of nuclear waste which has already been created is not known. This is because the amount of nuclear waste generated by the military is shrouded by uncertainty. What is known is that a typical 1,000 megawatt light water nuclear reactor will generate 200 to 350 cubic metres of low and intermediate level waste per year. It will also create about 75 square metres of high-level waste. This waste can be reduced in size, if it is reprocessed to less than half of what it started out as. The main component in such waste is caesium-137 and strontium-90. From the amount of nuclear power generation facilities in existence, each year, perhaps 200,000 cubic metres of low and intermediate waste and 10,000 square metres of high-level waste are created.[155]

As of 2014, there are 437 nuclear power reactors operating on Earth. Most projections suggest that, between 2015 and 2030, nuclear power could grow by a third. Whilst there is expected to be steady growth in Europe and North America, China, India and Russia currently have over 60 new reactors in construction. China plans to go from 20 reactors to 230 within the next 15 years. With this increase in nuclear energy, there will be an associated increase in the amounts of nuclear waste added each year. These increases in production will eclipse the current storage facilities. For example, in the United States, the passed over Yucca mountain storage facility was destined to hold 70,000 tons of high-level nuclear waste, yet by 2030, projections are that over 110,000 tons of high-level waste will need to be dealt with. Russia is understood to already have at least 140,000 tons of radioactive waste stored, whilst in the United Kingdom, the figure could be double this. A further contributing factor to the amount of waste is that produced from dismantling defunct reactors. A standard design, once demolished, will produce more than 100,000 tons of waste, including concrete, the reactor vessel, control rods and pipes. So far, only 17 of the world's 135 defunct reactors have been dismantled. In total, estimates suggest that whilst some 350,000 tons of spent nuclear fuel existed in 2013, by 2030, the figure could easily be double this, at 700,000 tons.[156]

---

[155] IAEA (2013). *Nuclear Waste*. (IAEA, Vienna). 2; Alvarez, R. (2011). 'Plutonium Wastes from the U.S. Nuclear Weapons Complex'. *Science & Global Security*. 19: 15–27; Dawson, F. (1997). 'An Overview of Military Radioactive Waste Management'. *Nuclear Engineering and Design*. 176: 23–6.

[156] Anon. (2013). 'A History of Fission'. *New Scientist*. 1 June. iii; Mascarelli, A. (2013). 'Eternal Challenge'. *New Scientist*. 2 Nov. 42–5; Anon. (2013). 'Nuclear Waste'. *New Scientist*. 1 June. iv; Schaffer, M. (2011). 'Toward a Viable Nuclear Waste Disposal Program'. *Energy Policy*. 39: 1382–8; Hou, Z. (2011). 'Government Policy and Future Projection for Nuclear Power in China'. *Journal*

The disposal of such nuclear waste into the oceans of the world was the clear pattern of the last half of the 20th century. Although the contribution of nuclear waste placed into the ocean is less than 0.1 percent of the overall radiation budget of the planet, this contribution is still significant. In terms of causation, the sinking of four nuclear submarines prior to the 1970s (two American, two Soviet) was probably the greatest contribution. This was followed by the dumping of low-level waste into the ocean by vessel, and then, the dumping from reprocessing plants. At the global level, the impacts on the human population of such nuclear waste streams into the ocean have been minimal. At local levels, the impacts of elevated levels of nuclear waste pollution in the ocean have been the subject of dispute.[157]

There have been three distinct sources of radioactive waste placed into the ocean. The first source has been reprocessing plants, of which three nuclear fuel reprocessing plants have resulted in significant direct releases into the oceans. They are the Sellafield plant on the Irish Sea on the north-west coast of England, the La Hague facility on the north-west coast of France and the plant at Trombay on the west coast of India. The most significant of these three has been Sellafield, which began discharging nuclear waste into the ocean in 1952. The peak of the exposure from this plant was in 1976 when dose rates to consumers of fish and shellfish who lived near Sellafield peaked at about 2 sieverts per year of which about 30 percent was due to actinides. The collective doses for all nuclides from Sellafield peaked in 1979, at 130 sieverts (of which the actinide contributions were less than 4.5 percent). By the early 1990s, following the introduction of new pollution control technologies, overall doses had reduced to about 0.1 sieverts per year, before continuing to fall during the following decade. Although considerable amounts of this waste continues to reside in the sediment and is taken up by local species, much of it has been dispersed. The perceived wisdom is that,

*of Energy Engineering*. Sept. 10–27. Anon. (2008). 'Nuclear Power's Great Expectation'. *Power Engineering International*. 16(9): 34, 36; Wald, M. (2007). 'What Now?' *Scientific American*. Aug. 48–56; Anon. (2006). 'Bury Waste, Again'. *New Scientist*. 6 May. 6.

157 Livingston, H. (2000). 'Anthropogenic Marine Radioactivity'. *Ocean & Coastal Management*. 43: 689–712; Solomon, K. (1998). 'Sources of Radioactivity in the Ocean Environment'. *Journal of Hazardous Materials*. 18(8): 255–62; Baxter, M. (1998). 'High-Activity Nuclear Wastes in the Oceans'. *Marine Pollution Bulletin*. 14(4): 126–32; Hamilton, E. (1998). 'Marine Environmental Radioactivity: The Missing Science?'. *Marine Pollution Bulletin*. 36(1): 8–18; Cormack, D. (1992). 'Scientific and Technical Background'. *Marine Policy*. Jan. 3, 11.

especially post-1990, food safety of any local marine species in this area has not been compromised.[158]

The second major source of nuclear waste being deposited into the ocean has been through dumping from vessels. This process began in 1946 at a site in the north-east Pacific Ocean, about 80 kilometres off the coast of California. The last intentional dumping of nuclear waste into the ocean was in 1993. In all, a total of 85 petabecquerels (PBq) of radioactive wastes were dumped in the Atlantic (45 PBq), the Pacific (1.4 PBq) and the Arctic (38 PBq). During the 48 year history of sea disposal, 14 countries used more than 80 sites to dispose of this radioactive waste – most of which was classified as low level. However, it is difficult to be certain about this as record keeping in the early decades was patchy.[159]

Between 1949 and 1982 several West European countries (principally the United Kingdom, but also including Belgium, the Netherlands and Switzerland) dumped about 54,000 terabecquerels (TBq) of radioactivity in some 140,000 tons of packaged waste at ten sites in the north-east Atlantic. The Atlantic dump sites made up 53.20 percent of the global activity in this area. The United Kingdom was responsible for 35,087 TBq of this total. Between 1946 and 1967, the United States dumped approximately 4,000 TBq of radioactive waste in about 90,000 containers

---

158 Alley, W. (2011). *Too Hot to Touch: The Problem of High Level Nuclear Waste*. (Cambridge UP, Cambridge). 36–50; Finegan, P. (2009). 'Accumulation, Solid Partitioning and Remobilisation of 99Tc in Subtidal and Intertidal Sediments in the Irish Sea'. *Continental Shelf Research*. 29: 1995–2010; Ould, Z. (2002). 'Assessment of Prospective Food-Chain Doses from Radioactive Discharges from Sellafield'. *Journal of Environmental Radioactivity*. 59: 273–91; Hunt, G. (1999). 'The Radiological Impact of Actinides Discharged to the Irish Sea'. *Journal of Environmental Radioactivity*. 44: 389–403; Berrow, S. (1998). 'Radionuclides in Harbour Porpoises *Phocoena phocoena* from British and Irish Coastal Waters'. *Marine Pollution Bulletin*. 36(8): 569–76; Leonard, K. (1997). 'Distribution of Technetium-99 in UK Coastal Waters'. *Marine Pollution Bulletin*. 34(8): 628–36; McCartney, A. (1994). 'Artificial Radionuclides in the Surface Sediments of the Irish Sea, 1968–1988'. *The Science of the Total Environment*. 141: 103–38; Jefferies, D. (1989). 'Observed and Predicted Concentrations of Caesium-137 in the Irish Sea 1970–1985'. *Journal of Environmental Radioactivity*. 10: 173–89.

159 Hamblin, J. (2009). *Poison in the Well: Radioactive Waste in the Oceans*. (Rutgers UP, New Jersey). 32–5, 46–51, 120–25; Povinec, P. (2000). 'AMS Measurements of 14C and 129I in Seawater Around Radioactive Waste Dump Sites'. *Nuclear Instruments and Methods in Physics Research B*. 172: 672–8.

of various types (although the 55 gallon steel drum was most popular) in the Pacific and Atlantic Oceans, and the Gulf of Mexico. With the Arctic sites, the former Soviet Union was responsible for 38,369 TBq (making them responsible for 45.10 percent of the global total); whilst in the Pacific sites, the Soviet Union was again at the forefront, but with a much lower contribution of 873.6 TBq (adding 1.01 percent to the global total). The United States contributed 554 TBq in the Pacific – or 0.66 percent of the global total. In terms of the types of waste dumped, close to 53.4 percent of the activity in the disposed radioactive waste was associated with the disposal of low-level packaged solid waste, of which some 93.5 percent was disposed of at the north-east Atlantic dumping sites by eight countries, mainly the United Kingdom. The dumping of low-level liquid and solid waste in the Arctic Ocean makes up less than 1.6 percent of the total activity dumped. The inventory of waste dumped into the Pacific Ocean amounts to close to 1.7 percent of the total activity dumped. Of all three of these oceans, only two of the sites in the north-east Atlantic have small increases in the measurement of carbon 14.[160]

The final notable source of nuclear waste being placed into the ocean involved covert disposal. The background to this problem was the former Soviet Union, which was emphatic during the 1980s that it 'had no intention of disposing of decommissioned nuclear powered submarines at sea'.[161] In reality, the Soviets' were emphatic liars. Between 1959 and 1992, the Soviet Union had dumped large amounts of radioactive wastes into rivers and seas. In addition to contaminated effluent and packaging waste, it dumped 16 nuclear reactors from submarines and icebreakers, some still with nuclear fuel, most of them in water less than 100 metres deep. These practices became apparent in 1991, when information was

---

160 Povinec, P. (2000). 'AMS Measurements of 14C and 129I in Seawater Around Radioactive Waste Dump Sites'. *Nuclear Instruments and Methods in Physics Research B*. 172: 672–8; IAEA (1999). *Inventory of Radioactive Waste Disposals at Sea*. (IAEA, Vienna). IAEA-TECDOC-1105; Edwards, R. (1995). 'Leaky Drums Spill Plutonium on Ocean Floor'. *New Scientist*. 22 July. 5; Burroughs, R. (1988). 'Ocean Dumping: Information and Policy Development in the United States'. *Marine Policy*. 99, 100; Mitchell, P. (1988). 'Gamma Radioactivity in the Iberian Marine Environment Closest to the NEA Dumping Site'. *Journal of Environmental Radioactivity*. 6: 77–89, Bewers, J. (1987). 'Analysis of the Issues Related to Sea Dumping of Radioactive Wastes'. *Marine Policy*. 105–20; Wastler, T. (1981). 'Ocean Dumping Permit Program under the London Dumping Convention'. *Chemosphere*. 10(6): 659–60.

161 International Maritime Organization. 'Report of the 12th Consultative Meeting of the Parties to the LDC'. London. 1 Dec. 1989. 34.

made public about the practice of the Soviet Union of dumping radioactive waste in the ocean around the Arctic. In 1992 the President of the Russian Federation convened a Commission on matters related to radioactive waste disposal at sea. The Commission's report suggests that the total amount of solid radioactive waste dumped into the northern Arctic seas was about 903 TBq, of which 0.0007 TBq was dumped into the Baltic, 3.7 TBq into the White Sea, 450 TBq into the Barents Sea and 315 TBq into the Kara Sea. The composition of these wastes included small ships, barges, packaged wastes and redundant equipment. Liquid radioactive wastes discharged into the Barents and Kara Seas and the Ara Bay were about 651 TBq. The quantity of spent nuclear fuel (predominantly fission products and actinides), expressed as strontium-90 equivalents, was about 1,704 TBq. This material was dumped in protected packages and in nuclear submarines.[162] Subsequent, independent, assessment showed that the dumping of this nuclear waste into the ocean contributed to the inventory of radionuclides. Specific increases were found with caesium-137 (1 PBq), strontium-90 (0.95 PBq), nickel-63 (0.3 PBq), cobalt-60 (0.1 PBq), plutonium-241 (78 TBq) and plutonium-240 (8.9 TBq). Despite these increases, the evidence suggested that the impacts of the dumped objects were small and localised to the immediate vicinity of the disposal sites. Overall, the levels of artificial radionuclides in the Kara and Barents Seas were considered negligible when compared with other natural sources, from which only insignificant impacts on the biodiversity and humans could be expected.[163]

---

[162] Hamblin, J. (2009). *Poison in the Well: Radioactive Waste in the Oceans.* (Rutgers UP, New Jersey). 2–4, 132–3, 178–80; Motavalli, J. (1998). 'Nuclear Waste: A Watery Grave?'. *E: The Environmental Magazine.* 9(1): 34; Templeton, D. (1997). 'Research Needed Relative to Radiological Assessment of the Dumping of Radioactive Wastes in the Oceans'. *Marine Pollution Bulletin.* 35(7): 374–80; MacKenzie, D. (1994). 'Doubts Lurk in Graveyard for Nuclear Subs'. *New Scientist.* 12 Mar. 4–5; Anon. (1993). 'Row over Sunken Nuclear Sub'. *New Scientist.* 18 Sept. 7; Anon. (1993). 'Whole Reactor Lurks under Barents Sea'. *New Scientist.* 13 Feb. 9; MacKenzie, D. (1993). 'Russia Owns Up to Sea Burial for Nuclear Waste'. *New Scientist.* 17 Apr. 5.

[163] The reported lines can be found in the 19th Meeting of the Parties to the LDC (1997). LC 19/10. Paragraph 6.9. For the IAEA report and associated commentary, see Sjoblom, K. (1999). 'International Arctic Seas Assessment Project'. *Science of the Total Environment.* 237: 153–66; Baxter, M. (1998). 'Modelling the Potential Radiological Consequences of Radioactive Waste Dumping in the Kara Sea'. *Journal of Environmental Radioactivity.* 39(2): 161–81; Champ, M. (1998). 'Assessment of the Impact of Nuclear Wastes in the Russian Arctic'. *Marine Pollution Bulletin.* 35(7): 203–21; Povinec, P. (1997).

## F. Industrial Wastes

The largest waste streams in modern society come from industrial sources. Unlike conventional common or municipal wastes, this is often collected and dealt with by the private sector, and is subject to its own regulations. Many hazardous wastes are industrial wastes. Historically, large amounts of industrial waste were disposed of in the oceans. This practice, which began in the 1940s, involved millions of tons of waste going into the sea. Most of this industrial waste (58 percent) was unwanted acids, 12 percent was refinery wastes, 7 percent was pesticides, 3 percent was paper mill waste and 20 percent was 'other'. It was estimated that, in 1959, the United States dumped around 2.2 million tons of this waste stream into the ocean, whilst by 1968 it had risen to 4.7 million tons. This trend continued during the 1970s. Similar practices were also common with other countries. Britain dumped 6 to 7 million tons of chemical waste, inert colliery waste, and fly ash between 1970 and 1980 into the ocean. In the early 1990s, Japan was the largest dumper of industrial wastes in the world, at around 4.5 million tons per year.[164] How much waste of similar types or quantities was being disposed of into the oceans by the developing countries is a matter of great uncertainty.[165]

---

'Summary of IAEA-MEL's Investigation of Kara Sea Radioactivity and Radiological Assessment'. *Marine Pollution Bulletin*. 35(7): 235–41; Carroll, J. (1997). 'Radionuclide Adsorption to Sediments from Nuclear Waste Dumping Sites in the Kara Sea'. *Marine Pollution Bulletin*. 35(7): 296–304; Vartanov, R. (1997). 'Nuclear Legacy of the Cold War: Russian Policy and the Cold War'. *Marine Policy*. 21(1): 1–15; Sazykina, A. (1997). 'Current and Potential Doses from Arctic Seafood Consumption'. *The Science of the Total Environment*. 202: 57–65; Baxter, M. (1995). 'Observations on Plutonium in the Oceans'. *Applied Radiation Isotopes*. 46(11): 1213–23.

[164] Graham, J. (2001). 'Japan and the Ocean Dumping of Wastes'. *Asian Studies Review*. 25(3): 335, 353–5; IMO (1991). *The London Dumping Convention: The First Decade*. (IMO, London). 4–6, 17–25; Kechum, B. (1981). *Ocean Dumping of Industrial Waste*. (Plenum, NYC). 7–10, 13–20; Duedall, I. (1983). *Wastes in the Ocean: Industrial and Sewage*. (Wiley, NYC). 5–17; Wastkerm T, (1981). 'Ocean Dumping under the London Dumping Convention in the United States'. *Chemosphere*. 10(6): 669–81; Wastler, T. (1981). 'Ocean Dumping Permit Program under the London Dumping Convention'. *Chemosphere*. 10(6): 659–68.

[165] Keyuan, Z. (2009). 'Regulation of Waste Dumping at Sea: The Chinese Practice'. *Ocean and Coastal Management*. 52: 383–9; IMO (1995). *Global Waste Survey*. (IMO, London). 2; Cormack, D. (1992). 'Scientific and Technical Background'. *Marine Policy*. Jan. 3, 9.

As discussed in forthcoming chapters, industrial waste is no longer disposed of in oceans. Rather, it is most commonly dealt with on land. When such wastes have no or only limited hazardous potential, such as uncontaminated soil and material associated with construction, or non-hazardous material used in farming, forestry or biomass, their management is often quite easy. However, when the industrial wastes have clear hazardous potential, much greater regulatory interest is generated. This is particularly obvious with the industrial wastes such as those from oil and gas, mineral processing and fossil fuel combustion industries. Of the latter, coal combustion wastes are particularly notable. For example, in the United States alone, more than 130 million tons of coal combustion waste is produced as coal is burnt for energy. Globally, the figure is probably around 750 million tons per year. Large amounts of this waste are either stored in stockpiles, disposed of in special ash landfills or lagooned. Although considerable amounts of work have gone into the reuse of such products, unless carefully stored, this waste represents a major potential source of release of many environmentally sensitive elements into the environment.[166]

A very similar situation exists with mining wastes. The difference is that mining wastes, by weight and size, are the largest waste stream on the planet. This is because a considerable amount of effort and waste is created to extract very small amounts of valuable resources. For example, a gold wedding ring containing 5 grams of gold would often leave 3 tons of waste. In the United States, mining generates nearly half of the total waste. In Europe, 28.3 percent of total wastes generated (at over 727 million tons) comes from waste material from mining and quarrying. Collectively, it has been estimated that the production of mine wastes is about several thousand million tons per year.[167]

Such wastes are often heterogeneous materials consisting of ore, gangue, industrial minerals, metals, coal or mineral fuels, rock, loose

---

[166] Riberio, J. (2014). 'Fly Ash from Coal Combustion: An Environmental Source of Organic Compounds'. *Applied Geochemistry*. 44: 234–54; Izquierdo, M. (2012). 'Leaching Behaviour of Elements from Coal Combustion Fly Ash: An Overview'. *International Journal of Coal Geology*. 94(5): 78–89; Kalembkiewicz, J. (2012). 'Ashes from Co-Combustion of Coal and Biomass'. *Resources Conservation and Recycling*. 69(12): 178–92.

[167] Note, in Europe, for 2010, construction waste accounted for 33.3 percent of the total – a greater amount than mining – at 855 million tons. Eurostat (2012). *Waste Statistics*. (EC, Brussels). 4; MacBride, S. (2012). *Recycling Reconsidered*. (MIT Press, Boston). 87–104; Hudson, K. (2011). 'Mine Wastes: Past, Present, Future'. *Elements*. 7: 375–80; Plumlee, D. (2011). 'Mine Wastes and Human Health'. *Elements*. 7: 399–405.

sediment, mill tailings, metallurgical slag and wastes, roasted ore, flue dust, ash and chemicals added to help the process which may include arsenic, lead and cyanide. Such waste can take up a great deal of space, blight the landscape and detrimentally impact upon local habitats as hazards – even if well managed. If the management turns poor, contaminated dusts may circulate, dams built to contain waste may be breached and mine tailings that leach may cause difficulties, with economic costs amounting to tens of billions of dollars.[168]

The need to deal with mining waste in domestic settings has been evident since the 19th century. In the 21st century, this is slowly emerging as an international concern. The global community has provided specific guidance for mines operating in internationally protected areas, such as World Heritage Sites,[169] as well as reminding states at the Rio+20 conference in 2012 that mining needed to be 'managed effectively and properly'.[170] Despite such clear needs, and it being the dominant source of most waste streams, mining waste is often excluded from generic waste management regimes. For example, in the United States, the 1976 Resource Conservation and Recovery Act (RCRA) was amended in 1986 to exclude 'solid waste from the extraction, beneficiation, and processing of ores and minerals' from regulation as hazardous waste under the RCRA.[171] This is not to suggest that no rules apply to mining waste, but rather, that different rules apply. Thus, whilst some regions such as Europe have quite full regulations, large parts of the developing world are silent in this area.[172]

---

168 MacBride, S. (2012). *Recycling Reconsidered.* (MIT Press, Boston). 87–104; Hudson, K. (2011). 'Mine Wastes: Past, Present, Future'. *Elements.* 7: 375–80; Plumlee, D. (2011). 'Mine Wastes and Human Health'. *Elements.* 7: 399–405.

169 See Gillespie, A. (2011). *Conservation, Biodiversity and International Law.* (Edward Elgar Publishing, Cheltenham and Northampton). Chapter 12.

170 Rio+20. A/CONF.216/XX. Paragraphs 227 and 228. For the earlier national approaches on this issue, see Lawrence, S. (2014). 'The Sludge Question: The Regulation of Mine Tailings in the Nineteenth Century'. *Environment and History.* 20(3): 385–410.

171 The RCRA was amended by adding section 3001(b)(3)(A)(ii). See Barringer, G. (2003). 'The RCRA Bevill Amendment: A Lasting Relief for Mining Wastes?' *Natural Resources & Environment.* 17(3): 155–75.

172 For the European rules, see Directive 2006/21/EC (15 Mar. 2006) on the Management of Waste from Extractive Industries. For best practice in this area, see Franks, D. (2011). 'Sustainable Development Principles for the Disposal of Mining and Mineral Processing Wastes'. *Resources Policy.* 36(2): 114–22; Mining, Minerals and Sustainable Development (2002). *Breaking New Ground.*

Similar considerations of being a dominant waste source (in terms of volume and weight) exist for waste from dredging waterways. Whilst mining is primarily done to extract economically valuable minerals, dredging is usually done to keep vessel-going waterways open for the purposes of navigation, to enlarge or deepen existing channel and port areas and to create new ones, or for engineering purposes, such as the creation of tunnels or laying of pipes. Due to the nature of the marine environment, these areas often have to be re-dredged on a regular basis, as otherwise they may fill up. Typically, the dredged materials are sediments which consist of alluvial deposits, that is, boulder, sand and mud. Often these are neither polluted nor cause bad environmental impacts when dumped. However, in many instances, what is dredged may contain toxic chemicals from land-based, or vessel-based, sources. These contaminated dredgings may cause environmental problems. In addition, these problems may be multiplied depending on where they are dumped. In many instances, this material may be so badly contaminated that it cannot be re-disposed into the ocean, and has to be disposed of on land. Even with relatively uncontaminated dredgings, if they are taken, or placed, indiscriminately, they may damage valuable coastal areas, such as surf breaks. Either way, dredging materials make up over 90 percent of all the total material dumped into the ocean. The vast majority of countries with coastlines are engaged in this activity to one degree or another, although it is not possible to be specific about how much waste is involved. Only a few countries have accurate reports in this area, such as with the United States, which dredged just under 54 million cubic metres annually in the first years of the 21st century. Unlike the other dominant streams of industrial waste, this is one area where the international community, through the London Dumping Convention, has provided advice on how it should be dealt with.[173]

---

(Earthscan, London). 230–40; Porter, R. (2002). *The Economics of Waste.* (Resources for the Future, Washington). 201.

[173] Monge, M. (2013). 'Morphodynamic Consequences of Dredging and Dumping'. *Ocean and Coastal Management.* 77: 40–48; Jeuken, M. (2010). 'The Impact of Dredging and Dumping on the Stability of Ebb-Flood Channel Systems'. *Coastal Engineering.* 57(6): 553–68; Lee, D. (2010). 'Scoping the Effective Marine Environmental Assessment of Dredging and Ocean Disposal'. *Marine Policy.* 34(5): 371–85; Moran, R. (2009). 'Out of Sight, Out of Mind: Ocean Dumping of Marine Wastes'. *World Watch.* 22(2): 30–34; Melnick, R. (2008). 'Dredging: Making Waves for Commerce or Environmental Destruction'. *Villanova Environmental Law Journal.* 19(1): 145–62; Hedge, L. (2009). 'Dredging Related Metal Bioaccumulation in Oysters'. *Marine Pollution Bulletin.* 58(6): 45–57; Blair, G. (1997). 'Dredging, Environmental Issues and Port Experience in

## 7. CONCLUSION

There are five relative certainties on waste. The first is what it is, that is, it involves solid material that is heavier than air, that it is unwanted and has no further use. The second certainty is that as the disposal of waste increases, the trend is to find the cheapest way to dispose of it. If those seeking to dispose of the waste have scruples and wish to follow the law (as opposed to dumping it illegally), the trend, as it has been for the last 200 years, is to dispose of it where there will be least resistance, and thus, lowest economic cost. Historically, this meant disposing of waste in either the commons, such as the ocean or rivers, or near communities without the capacity to object to the dumping. Those communities have always been those which are economically weak. Historically, this consideration of economic weakness was found within a country. Now it is also found globally, and as a third certainty, international law facilitates and attempts to regulate part of this trade. The reaction to this problem has been the creation of a new type of social movement, known as NIMBY. These movements, also, have become international and are no longer the monopoly of the working class in wealthy countries because now, as a broad rule, international law facilitates, and does not prohibit, the trade in hazardous waste for final disposal, if the importing countries clearly consent to the waste entering into their country and can safely deal with it. Although the practice is becoming clear that where a waste is clearly hazardous it should only be shipped from the developing world to the developed, this is by no means absolute. Accordingly, yet more waste may be dealt with inappropriately in the developing world.

Following on from the recognition that waste often involves social, economic and environmental costs, the fourth certainty arises – that individuals, private enterprises and policy makers want to reduce waste. No one is suggesting that more waste is a good thing. Rather, the clear consensus is that less waste should be created. The goals to slow the growth of waste are commonly found in local, national and international settings.

The fifth and final certainty with regard to waste is that the amounts of it being produced are increasing. Although some waste streams have stopped because of technological change, most waste streams have both diversified and increased. Each of the new types of waste requires a

the United States'. *Maritime Policy and Management*. 24(4): 313–18; Rasheed, K. (2001). 'Environmental Impacts of Dredging'. *International Journal of Environmental Studies*. 58(61): 703–15.

greater degree of management than the traditional types. The difficulties that this entails have been enhanced by increases in human numbers and per-capita rates of consumption for everything from consumer goods to more energy. No country has yet reached a point where its production of waste has begun to decline, although for a few high income countries, the rate of growth has slowed considerably in some waste streams. The rate of waste growth in the overwhelming majority of other countries, and other waste streams, continues to increase. These increases, both individually and collectively, represent one of the most obvious challenges for sustainability in the 21st century.

# 3. The waste hierarchy

## 1. INTRODUCTION

Quantities of waste are increasing exponentially in all parts of the world. Even in the most highly developed countries, with the best possible recycling schemes, waste production is continuing to grow, albeit at a smaller rate than elsewhere. No one wants more waste. The economic, social and environmental costs of this problem have resulted in clear goals, from the local to the international, to confront the problem. The question is, how?

The answer to this problem is the 'waste hierarchy'. The two broad aims of the hierarchy are to generate as little waste as possible in the first place, and then to extract the maximum practical benefit of the waste that is still produced. In terms of sequence, the hierarchy suggests that as a priority order in waste prevention and management legislation and policy, policy makers should pursue prevention; reuse; recycling; other recovery (such as burning it and capturing some value of the waste as an energy source); and then – as a last resort – disposal.

This chapter deals with the first part of the hierarchy – trying to reduce the production of waste in the first place. In this area, three options are available. First, the product can be prohibited. Second, consumers may voluntarily decide to steer away from the particular stream of waste, or third, pricing mechanisms are placed around it to make it economically unattractive, and thus create an incentive not to create the waste.

Finally, the producers of the product can be brought into the equation via ideals of 'extended producer responsibility', whereby they become responsible for dealing with the item that has ended up as waste, even if it has been owned privately in the interim years between the manufacture and final disposal. This last option is growing in popularity and is utilised in over 25 countries for a multitude of products ranging from consumer batteries to motor vehicles. Through such schemes of extended responsibility, those who control the design of such products are now being given a direct incentive to reduce the waste associated with the product, as they are now responsible for that too.

## 2. THE SOLID WASTE HIERARCHY

Although there is a difference between those seeking to minimise waste and those who wish to take it to zero, both are wedded to the same conceptual mechanism to achieve their goals. This is known as the 'solid waste hierarchy'.[1] This hierarchy has been adopted in numerous jurisdictions, with Europe and the United States being most notable.[2]

The hierarchy seeks to see waste in a broad sense, recognising that an overall strategy is required if the goal of confronting the waste problem is to be successful. Within the hierarchy there are two broad goals. The first is to generate as little waste as possible that has to be dealt with, and the second, to extract the maximum practical benefit of any waste that is still created, despite the efforts around the first goal. The overall intention is to allow as little waste as possible to go to final disposal in places such as landfill. The hierarchy, which applies in priority order, goes:

(a) prevention;
(b) preparing for reuse;
(c) recycling;
(d) other recovery, for example, energy recovery; and
(e) disposal.[3]

The waste hierarchy is far from absolute, and it has many theoretical difficulties in terms of being applicable to different places, different scales of operation and different types of waste. In many instances, deviation from the hierarchy may be the best thing to do.[4] Nevertheless,

---

[1] Bartl, A. (2011). 'Barriers Towards Achieving a Zero Waste Society'. *Waste Management.* 31: 2369–71; Rodrigo, N. (2009). 'The Waste Hierarchy Inverted Pyramid'. *Waste Management and Research.* 27(5): 407; Tjell, H. (2005). 'Is The "Waste Hierarchy" Sustainable?'. *Waste Management and Research.* 23: 173.

[2] Hultman, J. (2014). 'The European Waste Hierarchy: From the Sociomateriality of Waste to a Politics of Consumption'. *Environment and Planning A.* 44: 2413–27.

[3] Article 4. Directive 2008/98/EC (19 Nov. 2008) on Waste. Note, this diversion from the list may be justified in the case of the life-cycle thinking of a product. See generally, Hultman, J. (2012). 'The European Waste Hierarchy'. *Environmental and Planning A.* 44(1): 2413.

[4] Boyle, M. (2003). 'Scale and the Waste Management Hierarchy'. *European Planning Studies.* 11(4): 257–65; Price, J. (2000). 'Demand Management – A Basis for Waste Policy: A Critical Review of the Applicability of the Waste

as a broad set of principles based within a clear hierarchy, the ideas are both wise and well accepted in practice.

## 3. PREVENTION

### A. Product Prohibition

At the international level, prohibitions on products which go on to form particularly hazardous waste streams (that is, they were not prohibited because of their problem as wastes, but more so, as direct hazards) can be found for ozone depleting chemicals, persistent organic pollutants and mercury. Some regions also have restrictions on the use of heavy metals in some products.[5] The restrictions in all of these areas, although cognizant of waste impacts, were secondary considerations.

The exception to this approach at the global level, where a product has been restricted in terms of hazardous waste component, is with shipping. Here, under the 2009 International (Hong Kong) Convention for the Safe and Environmentally Sound Recycling of Ships, it was agreed that all ships, to be recycled in the future, must meet certain design and construction requirements. Most notably, it was agreed that they would prohibit and/or restrict the installation or use of listed hazardous materials, such as asbestos, ozone depleting chemicals, polychlorinated biphenyls (PCBs) and various anti-fouling compounds. Accordingly, in the future, it will be much safer to recycle this waste stream than previously.[6]

After the prohibitions on the products or processes which go on to form the most hazardous of wastes, there is a relative silence. In this area, such as dealing with more common forms of solid waste, it falls to each country, or region, to establish their own standards on which products should be prohibited or regulated. Many countries embarked on work in this area from the late 1970s, as consumer awareness began to focus on particular items of waste that could be easily prohibited by mandating that certain materials are used in their construction, or the product is prohibited in total. Debate in this area covered a multitude of consumer-related items such as disposable diapers, cigarette butts, polystyrene cups, plastic micro-beads in cosmetics and junk-mail/leaflets/flyers. However,

---

Hierarchy'. *Sustainable Development.* 8(2): 312–32; Barrett, A. (1997). 'Questioning the Waste Hierarchy'. *Journal of Environmental Planning and Management.* 40(1): 172–89.

[5] See Gillespie, A. (forthcoming). *Poison, Hazards and International Law.*

[6] The restricted products are found in Annex 1 of the 2009 Convention.

in all instances, despite clear public debate, actual restrictions on such products were few and far between beyond local and/or regional initiatives. Rather, the preferred approach in most national jurisdictions has been to attempt to control the problem through socially conscious consumerism, and if that was not sufficient, then by progressive pricing mechanisms.[7]

## B. The Green Consumer

Rather than prohibit certain consumer products because of their waste impacts, it has been hoped since the 1970s that the growth of the consumer movement could be harnessed to achieve the same goal, but without the same direct intervention in the market. Known as 'green consumerism', the ideal has developed that a citizen can influence environmental policy by their individual purchasing power. This market, worth tens of billions of dollars each year, has proved very successful, in both environmental and economic terms, around a series of key areas, such as food types, with exemplars being coffee, agriculture and fisheries. However, green consumerism has found itself to be limited in a number of areas. Fundamentally, it has typically become the preserve of the well educated and relatively affluent. Yet even this group will show fluctuating loyalty to the idea when the economic costs are too high, or they feel the information related to the product they are thinking of purchasing is not reliable.[8]

When it comes to the issue of the creation of waste and green consumerism, the link between consumer behaviour and the actual reduction of waste becomes even more tenuous. This is because, at base, green consumerism was never really intended for people to consume less,

---

[7] Novotny, T. (2014). 'Time to Kick Butts'. *New Scientist*. 28 June. 26; Anon. (2014). 'Plastic Crackdown'. *New Scientist*. 28 June. 7; Clark, B. (2007). 'London Councils Counter Freesheet Litter Problem'. *Marketing*. Feb. 21: 1.

[8] Akenji, L. (2014). 'Green Scapegoatism and the Limits to Green Consumerism'. *Journal of Cleaner Production*. 63: 315–37; Ubilava, D. (2010). 'Effects of Income and Social Awareness on Consumer Willingness to Pay'. *Technological Forecasting and Social Change*. 77(4): 587–93; Konstadinos, C. (2010). 'The Profile of the Green Consumer'. *International Journal of Consumer Studies*. 34(2): 153–60; Aldhous, P. (2010). 'Exposed: Green Shoppers Dirty Little Secret'. *New Scientist*. 27 Mar. 11; Aldhous, P. (2010). 'Hey Green Spender'. *New Scientist*. 20 Feb. 6–9; Porter, M. (1995). 'Green and Competitive'. *Harvard Business Review*. 73(5): 120–34; Anderson, W. (1972). 'The Socially Conscious Consumer'. *Journal of Marketing*. 36(3): 23–31.

but rather, to consume differently. It was (and is) more about changing patterns of consumption, rather than the volume of consumption.

This meant that environmentally conscious choices for, say food stuffs and making alternative choices about different sources of food has always been easier than actually deciding not to make the purchase at all and reducing the amount of material to dump. In addition to not being at the core of green consumerism, even with some of the more clearly defined areas where consumers could use their purchasing power to make an impact on waste reduction, say with disposable diapers, the impact has been minimal. Thus, for a chain like Starbucks, although their campaign for people to 'bring a cup' (and receive a ten cent discount) was taken up more than 26 million times in one year (and kept an estimated 1.2 million tons of paper out of landfills), the total amount of cups of coffee they sold overall in the year was 2.3 billion. This pattern is repeated in multiple other areas, with the same result being common – that despite the noble efforts of green consumers, they are never more than a small percentage of the overall market, and accordingly, in the best case scenarios, green consumerism can only reduce a specific waste stream by about 10 percent.[9]

In practice, this means that the waste continues to grow, and more forthright measures are required to deal with the problem, by trying to reduce the growth at source.

---

[9] Krausz, R. (2013). 'Zero Waste: A Reality Check'. *Waste Management*. May. 8–9; Aydinliyim, T. (2012). 'Reducing Packaging Waste and Cost via Consumer Price Discounts'. *Decision Sciences*. 43(6): 1063–74; World Bank (2012). *What a Waste: A Global Review of Solid Waste Management.* (World Bank, Urban Development Series, No. 15, Washington). 5; Pendry, L. (2012). Environmentally Friendly Parenting: Are Cloth Nappies a Step too Far?'. *Young Consumers*. 12(13): 5–19; Smith, J. (2011). 'Eco-Diaper System Wins Fans Quickly as Moms Awestruck to Find Cloth Diaper Option'. *Technology Business Journal*. 11: 567; Phillips, T. (2011). 'A Critical Review of a Key Waste Strategy Initiative in England: Zero Waste Places – Projects 2008–2009'. *Resources, Conservation and Recycling*. 55: 335–43; Cowlett, M. (2011). 'Do We Really Care About Sustainability?'. *Marketing*. 2: 32; Munasinghe, M. (2010). 'Can Sustainable Consumers and Producers Save the Planet?'. *Journal of Industrial Ecology*. 14(1): 4–6; Salhofer, S. (2008). 'Potentials for the Prevention of Municipal Solid Waste'. *Waste Management*. 28: 245–59; Stewart, B. (2004). 'Are We All Environmentalists Now? Rhetoric and Reality in Environmental Action'. *Geoforum*. 35(2): 231; Eriksson, C. (2004). 'Can Green Consumerism Replace Environmental Regulation?'. *Resource and Energy Economics*. 26(3): 281; Ebreo, A. (1999). 'Reducing Solid Waste: Linking Recycling to Environmentally Responsible Consumerism'. *Environment and Behaviour*. 31(1): 7.

## C. Progressive Pricing

If the waste reductions achieved by appealing to the conscientious consumer do not reduce enough waste, the next step is to create an economic mechanism which causes an incentive, or penalty, with regard to the production of waste. Most countries, especially in the developed world, have already done this by developing progressive pricing mechanisms, whereby some of the economic costs associated with the disposal of waste are cast back upon those who seek to dispose of it. These are often known as 'Pay-as-you-throw' mechanisms.[10]

These charges break with the tradition of paying for waste services through general blanket taxes or levies, and impose economic costs in direct proportion to the amount of waste produced. When the bill payers, especially those with less discretionary income, are also given other options to reduce their waste output and pursue legal options such as recycling (as opposed to illegal dumping) or purchase alternative products that do not produce as much waste, they will usually act to save money and divert as much of the weight of the waste as possible. In the developed world, such progressive pricing can reduce the amount of waste going into a landfill by between 20 and 30 percent.[11]

---

[10] Brown, Z. (2014). 'Better the Devil You Throw: Experience and Support for Pay as You Throw Waste Charges'. *Environmental Science and Policy*. 38: 132–44; Dahlen, L. (2010). 'Pay As You Throw'. *Waste Management*. 30(1): 293–309; Skumatz, L. (2008). 'Pay As You Throw in the U.S: Implementation, Impacts and Experience'. *Waste Management*. 28(12): 2778–89; Reichenbach, J. (2008). 'Status and Prospects of Pay As You Throw in Europe'. *Waste Management*. 28(12): 2801–14.

[11] Takeuchi, K. (2014). 'Evaluating Unit Based Pricing of Residential Solid Waste: A Panel Data Analysis'. *Environmental and Resource Economics*. 58(2): 245–72; Ino, I. (2011). 'Optimal Environmental Policy for Waste Disposal and Recycling when Firms are Not Compliant'. *Journal of Environmental Economics and Management*. 62: 290–308; Dresner, S. (2010). 'Charging for Domestic Waste: Environmental and Equity Considerations'. *Resources, Conservation and Recycling*. 54: 1100–108; Beukering, P. van (2009). 'Effectiveness of Unit-Based Pricing of Waste'. *Waste Management*. 29(11): 314–24; Skumatz, L. (2008). 'Pay As You Throw: Implementation, Impacts, and Experience'. *Waste Management*. 28: 2778–85; Bilitewski, B. (2008). 'From Traditional to Modern Fee Systems'. *Waste Management*. 28: 2760–66; Reichenbach, J. (2008). 'Status and Prospects of Pay-As-You-Throw'. *Waste Management*. 28: 2809–14; Sakai, S. (2008). 'Unit-Charging Programs for Municipal Solid Waste'. *Waste Management*. 28: 2815–25; Dunne, L. (2008). An Investigation into Waste Charges'. *Waste Management*. 28: 2826–34; McNeil, D. (2001). 'Hazardous Waste Disposal: A Comparative Analysis of Regulation, Taxes and Subsidies'. *Journal of Economics*

A similar result has occurred with progressive pricing on specific waste streams of strong public concern, such as disposable plastic bags. Here, the focus has been upon a product which, if not correctly dealt with, is a particularly disliked form of litter. The response of numerous cities in both developed and developing countries has been to either restrict the sale of plastic bags in general (and make alternative options such as paper bags easily available) or the type of bags that the plastic can be made out of, or more commonly, establish an additional cost on the bag. This additional cost then creates a disincentive to the taking of what was once free. Depending on the size of the additional cost, this then correlates with the size of the drop in their use. For example, in 2002 when Ireland began charging shoppers 0.22 euro per bag, within weeks the use of disposable plastic bags dropped by 94 percent.[12]

## D. Extended Producer Responsibility

The fourth option aimed at the prevention of waste is extended producer responsibility. The background context of this area is finding environmental solutions in cooperation with the private sector. This goal, well regarded at the Earth Summits in 2002[13] and 2012,[14] dovetails into the specific topic of trying to control waste.

Private enterprise is often interested in this area, as they do not wish their brand names and associated reputation to be associated with waste problems, and wish to be viewed as part of the solution, not the source,

---

*and Finance*. 24: 280–303; Choo, C. (1999). 'An Economic Analysis of Household Waste Management'. *Journal of Environmental Economics and Management*. 38: 234–46; Anderson, M. (1998). 'Assessing the Effectiveness of Waste Tax'. *Environment*. 40(4): 10–15; Palmer, K. (1997). 'The Cost of Reducing Municipal Solid Waste'. *Journal of Environmental Economics and Management*. 33: 128–50; Ackerman, F. (1992). 'Waste Management: Taxing the Trash Away'. *Environment*. 34(5): 2–10.

12 MacBride, S. (2012). *Recycling Reconsidered*. (MIT Press, Boston). 197–200; Haoran, H. (2012). 'Effects of Environmental Policy on Consumption: Lessons from the Chinese Plastic Bag Regulation'. *Environment and Development Economics*. 17(4): 303–20; Freinkel, S. (2011). *Plastic*. (Houghton, NYC). 141, 148, 151; Syeda, A. (2011). 'Plastic Waste Management Strategies for Indian Cities'. *Journal of Chemical, Biological and Physical Sciences*. 2(1): 40–57; Anon. (2005). 'Japan's Tax on Plastic Sacks'. *New Scientist*. 18 June. 4; Williams, C. (2004). 'Battle of the Bag'. *New Scientist*. 11 Sept. 30–33.

13 The 2002 World Summit on Sustainable Development. Plan of Implementation. Section. III: 22.

14 RIO+20. A/CONF.216/XX. Paragraphs 46, 69, 71.

to specific waste problems.[15] The first entrepreneur to recognise this was probably Henry Ford in the 1930s. Ford tried to replicate the success he had with assembly lines by establishing a large scale disassembly line. This involved hundreds of men deconstructing dozens of automobiles per day, back into their original components. The only problem was that the costs of the labour vastly outweighed the economic return on the deconstructed cars. Moreover, given there was no mandate that this had to be done, Ford eventually closed the plant, allowing the consumers of his cars to deal with them when they came to the end of their working life. Without any direction or incentive to act otherwise, the owners of the vehicles preferred the option of abandonment as the most common form of disposal.[16]

In the 21st century, this type of idea has been revitalised at both the national and international levels. Of the latter, under the auspices of the Basel Convention and drawing on their general principle that the generator of the waste should have an input into its final disposal,[17] the parties have gone on to establish partnerships with specific industrial sectors, with a view to promoting environmentally sound management of hazardous wastes and minimisation.[18]

The successes of the Basel Convention in this area have been most obvious with the Mobile Phone Partnership Initiative. Through this, 12 manufacturers and three telecom operators, in cooperation with the parties to, and bodies within, the Convention, have developed a series of tools. These include five technical guidelines which include awareness raising on design considerations, collection of used and end-of-life mobile phones, transboundary movement of collected mobile phones, refurbishment of used mobile phones, and material recovery/recycling of end-of-life mobile phones.[19] All of these guidelines are designed to help improve sustainable practices in this area.

A similar process followed suit with e-waste in general, and computer waste in particular, whereby the parties, in association with key industry players, have begun a process which aims to establish standards for the

---

[15] Roper, P. (2006). 'How (and Where) the Mighty Have Fallen: Branded Litter'. *Journal of Marketing Management*. 22: 473–87.

[16] Minter, A. (2013). *Junkyard Planet*. (Bloomsbury, NYC). 170.

[17] As noted in the preamble of the Basel Convention.

[18] For the related Basel Decisions on Partnerships with Industry in this area, see Decisions V/13 (1999); VI/32 (2002); VII/7, VII/3 (2004); VIII/5 (2006); IX/7 (2008); X/19 (2011); X/21 (2011); X/26 (2011).

[19] For the Decisions on the Mobile Phone Partnership, see VI/31 (2002); VII/4 (2004); VIII/6 (2006); and IX/8 (2008).

safe management of electronic waste, as well as life-cycle considerations in the overall design and operation of the products.[20]

Although these international partnerships are commendable, they do not involve any specific obligations upon the industries that create the product in the first place to possess a clear responsibility to deal with the waste that the product ends up as, in the last place. The only mechanism that makes this linkage is extended producer responsibility. The genesis of this scheme began in Germany in 1991, with their Green Dot programme. This required those who produced packaging waste to be responsible for it.[21] In practice, this meant consumers could leave packaging with retailers, and packagers were required to internalise the costs for its disposal and pay for the recycling and disposal costs.[22]

The Green Dot programme ultimately evolved into a much more comprehensive strategy that obliges the manufacturer of the product to be responsible for the entire life cycle of the product and especially for the take back, recycling and/or its final disposal. Accordingly, the producer's responsibility for a product is extended to the post-consumer stage of a product's life cycle, and therefore a substantial part of the costs of dealing appropriately with the waste are shared with those who made the original profit in the creation of the product in need of final disposal.[23]

---

20 For the Decisions on electronic waste/computer equipment, see IX/6 (2008); IX/9 (2008); and X/20 (2011).

21 'Regulation on the Avoidance of Packaging Waste'. 31 ILM (1992) 1135. Also, Nakajima, N. (2006). 'A Description and Analysis of the German Packaging Take-Back System'. *Bulletin of Science Technology & Society*. 26: 510–18.

22 Rotter, J. (2011). 'Waste Management and Producer Responsibility: A Score Behind – a New Ahead'. *Waste Management Resources*. 29: 889; Fleckinger, P. (2010). 'The Organization of Extended Producer Responsibility in Waste Policy'. *Journal of Environmental Economics and Management*. 59: 57–66; Khetriwal, D. (2009). 'Producer Responsibility for E-Waste Management'. *Journal of Environmental Management*. 90: 153–65; Nixon, H. (2007). 'Financing Electronic Waste Recycling: Californian Households' Willingness to Pay Advanced Recycling Fees'. *Journal of Environmental Management*. 84: 547–59; Gottberg, A. (2006). 'Producer Responsibility, Waste Minimisation and the WEEE Directive'. *Science of the Total Environment*. 359: 38–56; Porter, R. (2002). *The Economics of Waste*. (Resources for the Future, Washington). 33–5.

23 Austen, A. (2013). 'Where Will All the Waste Go? Utilitising Extended Producer Responsibility Framework Laws'. *Golden Gate University Environmental Law Journal*. 6(2): 178–89; Lifset, R. (2013). 'Extended Producer Responsibility'. *Journal of Industrial Ecology*. 17(2): 162–70; Gui, L. (2013). 'Implementing Extended Producer Responsibility'. *Journal of Industrial Ecology*. 17(2): 167–77; Dubois, M. (2012). 'Extended Producer Responsibility for

It is through the obligatory placement of the final burden and all of its associated economic costs on the original producer that it is hoped that innovative solutions will be found. Solutions should be generated here, as the original producers are the ones who are in the best position to control the longevity, content and recyclability of the products that they design and market.[24]

By 2014, extended producer responsibility, as supported by the OECD,[25] was found in over 400 schemes in operation across the OECD and developing countries.[26] For example, the 2008 Waste Minimisation Act of New Zealand allows their parliament, if convinced that a product will or may cause significant environmental harm when it becomes waste, or there are significant benefits from reduction, reuse, recycling, recovery or treatment of the product, to place the product under a product stewardship scheme. The goal of this is:

> [T]o encourage the people and organisations involved in the life of a product to share responsibility for ensuring there is effective reduction, reuse, recycling, or recovery of the product; and managing any environmental harm arising from the product when it becomes waste.[27]

Although extended producer responsibility schemes are now global, it is within Europe that they are most dominant. According to the 2008 European Directive on Waste, in order to strengthen the reuse and the prevention, recycling and other recovery of waste, member states may adopt measures to ensure that anyone who professionally develops,

---

Consumer Waste'. *Waste Management and Research.* 30(9): 408–24; Shinkuma, T. (2012). 'Effectiveness of Policy Against Illegal Disposal of Waste'. *Environmental Economics and Policy Studies.* 14(2): 298–315.

24 Coggins, C. (2001). 'Waste Prevention – an Issue of Shared Responsibility for UK Producers and Consumers: Policy Options and Measurement'. *Resources, Conservation and Recycling*. 32: 181–90.

25 OECD (2005). *Analytical Framework for Evaluating the Costs and Benefits of Extended Producer Responsibility Programmes* (OECD, Paris); OECD (2005). *EPR Policies and Product Design: Economic Theory and Selected Case Studies.* (OECD, Paris); OECD (2001). *Extended Producer Responsibility: A Guidance Manual for Governments.* (OECD, Paris).

26 See OECD (2014). 'Global Forum on the Environment: Promoting Sustainable Materials Management Through Extended Producer Responsibility'. Available from: www.oecd.org/environment/waste/Global-Forum-Co-chairs-Summary-June-2014.pdf, visited 26 Mar. 2015.

27 Article 8. Waste Minimisation Act. Public Act 2008. No. 89 (NZ). The basis for the consideration is found in article 9.

manufactures, processes, treats, sells or imports products (producer of the product) has extended producer responsibility. Specifically:

> Such measures may include an acceptance of returned products and of the waste that remains after those products have been used, as well as the subsequent management of the waste and financial responsibility for such activities. These measures may include the obligation to provide publicly available information as to the extent to which the product is re-usable and recyclable.

It was added that such measures could help encourage the design of products in order to reduce their environmental impacts and the generation of waste in the course of the production and subsequent use via making them both more durable, and once waste, easier to recycle or more effectively dispose of.[28]

From this thinking, producer responsibilities now exist for the improved product design (in terms of fewer pollutants, and improved ease of recycling), obligatory collection rates and environmentally safe disposal of the residue waste. Within Europe this obligation covers packaging,[29] automobiles,[30] batteries and electronic appliances. In particular, with electronic waste, the disposed-of product has to be kept separate from the general municipal waste stream and alternate disposal systems must be in place, free of charge for the consumer. From 2016, the minimum collection rate for all electronic waste must be above 45 percent. By 2019 it must be above 65 percent. This is supplemented by recovery targets for both reused and recycled (by weight of material collected), depending on the type of e-waste, ranging between 50 and 85 percent.[31]

Similarly, with the collection and recycling of consumer batteries within Europe, all member states have to maximise the separate collection and recycling of waste batteries (and accumulators) and minimise their disposal as mixed municipal waste in order to achieve a high level of recycling. This must be at no cost for the last purchaser of the working product. In terms of collection schemes, this means that each member

---

[28] Article 8. Directive 2008/98/EC (19 Nov. 2008) on Waste.

[29] Directive 94/62/EC (20 Dec. 1994) on Packaging and Packaging Waste. OJ L 365. 31 Dec. 1994. p. 10.

[30] Directive 2000/53/EC (18 Sept. 2000) on End-of-Life Vehicles.

[31] Articles 5, 6, 7, 8, 9, 11 and Annex V. Directive 2012/19/EU (4 July 2012) on Waste Electrical and Electronic Equipment. Note, the targets are calculated on the basis of the total weight of waste electronic equipment collected in a given year in the member state concerned/4 kilos of collected e-waste per person, per year.

state has to enable suitable collection points for the collection, and then recycling, of waste batteries. The recycling target, to be achieved by 2016, is 75 percent by average weight of nickel-cadmium batteries and accumulators, and 50 percent by average weight of all other waste batteries and accumulators.[32]

This scheme for the recycling of batteries, which has seen strong support by nearly 500 industries involved in the manufacture of batteries, has already made good progress in meeting its target. A large amount of this success has been down to the ease with which the collection is carried out. For example, in Britain, any shop or online retailers that sell more than 32 kilograms of batteries a year (the equivalent to one pack of 4 AA batteries a day) must offer facilities to collect the used batteries, which are then sent back to the manufacturer for the next step of either safe disposal or recycling. Similar models to this have been adopted in Australia and in North America, where the (non-governmental) Rechargeable Battery Take Back Coalition has collected and recycled (and never landfilled) more than 70 million rechargeable batteries, via consumers dropping them off for recycling at a large number of centres, and participating retail stores in particular, in both Canada and the United States.[33]

## 4. CONCLUSION

Quantities of waste are increasing everywhere. These wastes, more often than not, have economic, social and environmental costs. To confront this problem a waste hierarchy has been adopted. The two broad aims of the hierarchy are to generate as little waste as possible in the first place, and then to extract the maximum practical benefit of the waste that is still produced. In terms of sequence, the hierarchy suggests that, in priority, policy makers should pursue prevention; then reuse; recycling; other recovery (such as energy recovery); and then – as a last resort – disposal.

---

[32] See articles 8 and 10 of Directive 2006/66/EC (6 Sept. 2006) on Batteries and Accumulators. See also Annex III.

[33] MacBride, S. (2012). *Recycling Reconsidered.* (MIT Press, Boston). 188–9; Anon. (2011). 'Recharging Battery Stewardship'. *WME Magazine*. Mar. 7–10; Bernardes, A. (2003). 'Collection and Recycling of Portable Batteries'. *Journal of Power Sources*. 124 (2003): 586–92; Farhni, F. (1995). 'Battery-Recycling as a Part of the Swiss Waste Management Concept'. *Journal of Power Sources*. 57: 47–9.

This chapter dealt with the first part of the hierarchy – trying to reduce the production of waste in the first place. In this area, three options are available. First, the product can be prohibited. The problem here, especially when dealing with any waste that is not obviously hazardous, is that few legislators wish to actively prohibit consumer choice. Accordingly, a second method becomes available. This is that consumers may voluntarily decide to steer away from the particular stream of waste. The difficulty with this approach is that such voluntary action has never done more than slightly dent the growth, perhaps up to 10 percent, in any particular waste stream.

The third option is progressive pricing, whereby the disposal of waste generates a specific economic cost, thus making it unattractive to dump. If done correctly, and then there are adequate mechanisms to support recycling, alternative purchasing options, and the option for illegal dumping is controlled, this may make reductions in a waste stream of between 20 and 30 percent.

Finally, the producers of the product can be brought into the equation via ideals of extended producer responsibility, whereby those who initially design and manufacture the product also become responsible for dealing with the item when it finally becomes waste. This last option is growing in popularity with over 400 schemes, covering everything from consumer batteries to motor vehicles, in operation across the OECD and developing countries.

Ultimately, by themselves, none of these mechanisms will be able to prevent waste being generated in the first place. Although there is a clear limit in the willingness to prohibit consumer-related products that end up as wastes which are non-hazardous and consumer influence is ultimately small in this area, models which adopt both progressive pricing policies to deter waste, and schemes of extended producer responsibility, hold significant promise. This is especially so if their longer term targets of over 50 percent of specific waste streams are dealt with as expected, recycled correctly and diverted from the waste stream. In both instances, placing economic and legal burdens upon those who design the product, and on those further down the chain who may be considering how to dispose of it, are moving in the correct direction. Of course, neither of them will stop the production of waste, but they are the foremost weapons by which to slow its expansion, before considerations of disposal come into play.

# 4. Recycling

## 1. INTRODUCTION

This chapter is about recycling and related practices. The recycling symbol is one of the most easy to identify labels of environmental responsibility. In the 21st century, it appears that environmental, economic and social goals have combined to support recycling and make this practice the common form of environmental action for many, if not most, citizens. It is also actively pursued by many private enterprises and governments trying to tackle unwanted waste. The goals in this area are well entrenched in domestic, regional and international law and policy, with a number of specific international laws being directly targeted to address specific waste streams for recycling, such as old ships, or the overall trade in unwanted materials for recycling ranging from plastic to electronic waste. However, as experience has shown over the last three decades, at each point, the economic, social and environmental considerations related to recycling must be carefully considered.

## 2. RECYCLING, REUSE, RECOVERY AND RECLAMATION

The terms recycling, reuse, recovery and reclamation are often used interchangeably. However, each of these terms is very different, and their dissimilarities can make a mountain of difference with how a waste stream is dealt with. For example, if an old computer is to be reused, in most instances, it can be traded internationally as it is not hazardous waste. However, if it is to be recycled, it may be hazardous waste, and strong caution is urged in any potential international trade. Accordingly, great effort is put into mechanisms to distinguish between when old electronic material is waste or non-waste.[1]

---

[1] See Decision BC-11/4 (2013). Technical Guidelines on Transboundary Movements of Electronic and Electrical Waste, In Particular, Regarding the Distinction Between Waste and Non-Waste under the Basel Convention.

For the purposes of clarity, *recycling* means the reprocessing of waste or diverted material to produce *new* materials, products or substances.[2] To recycle glass means to melt the unwanted product down and remake it into a new glass product (about 30 percent of the market in most countries) or grind it up to be used as a substitute in a number of modern applications such as road building, insulation foam, or for gravel or sand in some concrete mixes. To recycle an old tyre is to shred, granulate or chemically take it apart and carefully feed it back into virgin compounds that make everything from rubber-based playgrounds to motorway surfaces (with the latter taking 12 million scrap tyres per year in the United States alone). In the case of used clothing given to charity each year, 45 percent is recycled and is turned into rags or respun to make fibre for industrial purposes.[3] A final example of recycling involves the use of biowaste (garden and park waste, food and kitchen waste from households, restaurants and retail premises and comparable waste from food processing plants)[4] into compost. Composting involves decomposing organic matter and recycling it as fertiliser and/or soil supplement.[5]

---

[2] This definition is from article 6 of New Zealand's Waste Minimisation Act. Public Act 2008, No. 89 (NZ).

[3] Vanguilder, C. (2012). *Hazardous Waste Management*. (Mercury, Boston). 61–2; MacBride, S. (2012). *Recycling Reconsidered*. (MIT Press, Boston). 126; Ejiogu, A. (2012). 'E-Waste Economics: A Nigerian Perspective'. *Management of Environmental Quality: An International Journal*. 24(2): 199–213; Anon. (2012). 'Vital for the Poor'. *Economist*. 10 Nov. 48; UNEP (2011). *Where Are WEee in Africa? E-Waste Africa Programme*. (UNEP, Nairobi). 3–4; Yoshida, A. (2010). 'Reuse of Secondhand TVs Exported from Japan'. *Waste Management*. 30: 1063–72; McNeil, J. (2008). 'The Afterlife of American Clothes'. *Reason*. 40(4): 68–9.

[4] This definition is from article 3 of Directive 2008/98/EC (19 Nov. 2008) on Waste.

[5] The options for composting range from simply layering rotting material on top of each other, through to closely monitored processes in which air, water, temperature and chemicals are added or removed to assist the process of decomposition. These options may include anaerobic or aerobic methods. The former are based around slow decomposition, usually sealed and underground, wet and without fresh oxygen. The latter are based around quicker decomposition, above ground, dry with access to air and often movement. For commentary in this area, see Anon. (2010). 'Managing Organic Municipal Waste'. *Appropriate Technology*. 37(3): 62–7; Domingo, J. (2009). 'Domestic Waste Composting Facilities: A Review of Human Health Risks'. *Environment International*. 35: 382–9; Nadal, J. (2009). 'Health Risks of Organic Waste'. *International Journal of Hygiene and Environmental Health*. 212: 661–9; Giusti, L. (2009). 'A Review of Waste Management Practices and Their Impact on Human Health'. *Waste*

*Reuse* means the further use of waste or diverted material in its existing form for the original purpose of the materials or products that constitute the waste or diverted material, or for a similar purpose.[6] Reuse is a method of diversion of discards where they are used, again, for a similar or same purpose. Used clothes, salvaged building materials, repaired appliances, retreaded tyres or refilled bottles are all examples of reuse, as opposed to being discarded after one cycle of use. Reusing such unwanted materials, which are still functional, has a strong benefit to people who could not afford them otherwise. For example, about 25 percent of all the hard-drives from computers sent to recycling depots in the United States have been used for less than 500 hours. For all effective purposes, these drives can be used for hours more, especially in places where the next consumer is more concerned about price rather than the age of the product. Thus, over 70 percent of the electronic material shipped to Africa ends up being reused or refurbished for further use. Similarly, between 30 and 35 percent of all tyres have more than one life in most industrialised countries after they are retreaded and reused. Likewise, of 7 billion pounds of used clothing that is given to charity each year in the United States and is then passed on to over 100 countries, 45 percent are reused clothes.[7]

*Recovery* involves operations, the principal result of which is that waste ends up serving a useful purpose by replacing other materials which would otherwise have been used to fulfil a particular function, or waste being prepared to fulfil that function, in the plant or in the wider

---

*Management*. 29: 2227–39; Gomes, A. (2008). 'Separate Collection of the Biodegradable Waste'. *Waste Management*. 28(10): 1711–19; Diaz, L. (2006). 'Managing the Organic Fraction of Municipal Solid Waste'. *BioCycle*. 47(10): 50–52; Strasser, S. (1999). *Waste and Want: A Social History of Trash*. (Holt, NYC). 153–5, 230, 259.

[6] This is the European definition of reuse, in Directive 2008/98/EC (19 Nov. 2008) on Waste. See also article 2(13). This definition is from article 6 of New Zealand's Waste Minimisation Act. Public Act 2008, No. 89 (NZ).

[7] Vanguilder, C. (2012). *Hazardous Waste Management*. (Mercury, Boston). 61–2; MacBride, S. (2012). *Recycling Reconsidered*. (MIT Press, Boston). 126; Ejiogu, A. (2012). 'E-Waste Economics: A Nigerian Perspective'. *Management of Environmental Quality: An International Journal*. 24(2): 199–213; Anon. (2012). 'Vital for the Poor'. *Economist*. 10 Nov. 48; UNEP (2011). *Where Are WEee in Africa? E-Waste Africa Programme*. (UNEP, Nairobi). 3–4; Farrant, L. (2010). 'The Environmental Benefits from Reusing Clothes'. *The International Journal of Life Cycle Assessment*. 15(7): 726–36; Yoshida, A. (2010). 'Reuse of Secondhand TVs Exported from Japan'. *Waste Management*. 30: 1063–72; McNeil, J. (2008). 'The Afterlife of American Clothes'. *Reason*. 40(4): 68–9.

economy.[8] The foremost example of recovery is the burning of waste for the creation and capture of energy, through which the value of the energy is recovered from the waste and then put to other purposes. Thus, tens of thousands of old tyres are burnt for energy recovery (a ton of tyres is the equivalent to a ton of good quality coal, or 0.7 ton of fuel oil), as are 10 percent of all old clothes given to charity. Similarly, old paper has a high value as fuel for incinerators.

Within industrial contexts, reuse and recovery often become known as *reclamation*, whereby waste is run through a process to recover a useful or valuable material for future use. Examples of other types of industrial reclamation are distilling waste solvent to reclaim the solvent, burning waste oil to recover the heat value, or reclaiming acids and bases from plating solutions to use them for further manufacturing processes. In the case of the international conventions designed to protect the ozone layer, reclamation has been defined as meaning the reprocessing and upgrading of recovered controlled ozone depleting substances through specific mechanisms, so that they can be used again for another controlled chemical process.[9]

## 3. LAW AND POLICY

The national recognition of the value of recycling in both law and policy is not new. The first formal attempts by governments to control this area, for reasons of security of resource supply, occurred in both the First and Second World Wars on both sides of the Atlantic. In the United States, this began in 1917 when *Scientific American* published a short article entitled 'Junk Is America's Richest War Bride: The Result of Amazing Wastefulness'. In a time of shortage of key raw materials, the contemporary, educated and patriotic readers knew what to do next. Accordingly, with strong government direction in both the United States and Britain (in both wars), nationally organised recycling efforts showed how large amounts of material (some 25 percent of each waste stream), especially metal, rubber and rags/paper, could be easily collected because of their strategic value to a nation. Similarly, the trade in metals for recycling, such as the 3.5 million tons of scrap steel going to Japan (in 1939), was suddenly stopped for reasons of national security. Thus, the collection

---

[8] This definition is taken from the European Directive 2008/98/EC (19 Nov. 2008) on Waste. See article 2(15).

[9] Decision IV/24. Recovery, reclamation and recycling of controlled substances.

and recycling of such valuable resources became the subject of strictly national, high profile, public campaigns. These campaigns were so successful that in the United States alone, by the end of the conflict, there were over 3,000 scrap iron and steel firms. However, when the conflicts stopped, so too did the active interest and state support for recycling. At the same time, the costs of virgin materials were very cheap, the costs of labour for recycling were expensive, the costs of dumping were non-existent and the demand for new, as opposed to recycled, products trumped everything. This meant that with products like motor vehicles, in the United States between 1950 and 1970, up to 40 million were just abandoned when they came to the end of their working life.[10]

Things began to change back in the 1960s. Specifically, the American Congress in their 1965 Solid Waste Disposal Act[11] found 'the failure or inability to salvage and reuse such materials economically results in the unnecessary waste and depletion of our natural resources'.[12] A little over a decade later, in their 1976 Resource Recovery Act, they noted:

> [M]illions of tons of recoverable material which could be used are needlessly buried each year; methods are available to separate usable materials from solid waste; and the recovery and conservation of such materials can reduce the dependence of the United States on foreign resources and reduce the deficit in its balance of payments.[13]

---

[10] Minter, A. (2013). *Junkyard Planet*. (Bloomsbury, NYC). 82–3, 162–4; Oldenziel, R. (2013). 'Reconsidering Recycling'. *Contemporary European History*. 22(3): 347–63; Matthew, E. (2011). 'Aluminium and the Environmental History of the Second World War'. *Environmental History*. 16(1): 69–93; Cooper, T. (2006). 'Rags, Bones and Recycling Bins'. *History Today*. 56(2): 17–18; Rogers, H. (2005). *Gone Tomorrow: The Hidden Life of Garbage*. (Norton, NYC). 134–5; Strasser, S. (1999). *Waste and Want: A Social History of Trash*. (Holt, NYC). 13–20.

[11] P.L. 89-272, 79 Stat. 992.

[12] Solid Waste Disposal Act. 79 Stat. 992 1965. Also, Public Law 89-272. 20 Oct. 1965. Section 202.

[13] Resource Recovery Act. 90 Stat. 2795 1976. Also, Public Law 94-580. 21 Oct. 1976. Section 1002(c). Note also that the 1990 Pollution Prevention Act recorded the general principle that 'pollution that cannot be prevented should be recycled in an environmentally safe manner whenever feasible'. Section 6602(b). See generally, Vanguilder, C. (2012). *Hazardous Waste Management*. (Mercury, Boston). 16–17.

Since this point, the United States, like many other countries, has adopted nationally, and/or locally, specific targets for recycling in particular streams of waste.[14]

The regional and international recognition of the value of recycling and reuse can be traced to the early 1990s. These moves have been actively encouraged by the OECD, who have argued that 'Member countries should … [in certain circumstances] provide incentives [from organising collection schemes to creating markets for certain recycled goods] to take part in … environmentally sound recycling schemes'.[15] In Europe, Member States are obliged to:

> [T]ake measures, as appropriate, to promote the re-use of products and preparing for re-use activities, notably by encouraging the establishment and support of re-use and repair networks, the use of economic instruments, procurement criteria, quantitative objectives or other measures. [They shall] … promote high quality recycling and, to this end, shall set up separate collections of waste where technically, environmentally and economically practicable and appropriate to meet the necessary quality standards for the relevant recycling sectors.[16]

A series of international gatherings have also given their support to the general need for recycling. The original hope in this area, as set down in 1992 in Agenda 21, was that all countries would 'strengthen and increase national … reuse and recycling systems'.[17] Specifically, 'by the year 2000, in all industrialized countries, and by the year 2010, in all developing countries, [all would] have a national programme, including, to the extent possible, targets for efficient … reuse and recycling'.[18] Twenty years later at the 2012 'Rio+20' conference, the participants added:

> We therefore commit to further reduce, reuse and recycle waste … with a view to managing the majority of global waste in an environmentally sound manner and, where possible, as a resource.[19]

---

[14] See for example, Hill, A. (2014). 'Modelling Recycling Targets'. *Journal of Environmental Protection*. 5(7): 627–36.

[15] OECD (2007). *Guidance Manual on Environmentally Sound Management of Waste* (OECD, Paris). Recommendation 9.

[16] Directive 2008/98/EC (19 Nov. 2008) on Waste. Article 11.

[17] Agenda 21 (1992). UNCED. Doc A/CONF.151/4. Chapter 21, paragraph 17.

[18] Agenda 21 (1992). UNCED. Doc A/CONF.151/4. Chapter 21, paragraph 18.

[19] RIO+20. A/CONF.216/XX. Paragraph 218.

In addition to these generic goals, a number of specific environmental regimes also directly address recycling. The first example is the 1987 Montreal Protocol on Substances that Deplete the Ozone Layer (the Montreal Protocol). This Protocol strictly controls the trade in ozone depleting chemicals, but in doing so, also facilitates their recycling or reclamation. The Parties have continually strived to develop the 'best technologies' and management systems in this area. They have also created 'banks' of used and extracted ozone depleting chemicals that can be reused. These chemicals are very important because they can be reclaimed and put into other approved products – without being seen as a breach of compliance with the Montreal Protocol.[20]

The second regime through which a similar process is likely to develop, akin to the Montreal Protocol, is the 2013 Minamata Convention on Mercury. This regime, due to the strong recycling options for mercury, has provisioned itself to recognise the use of 'recovered, recycled, reclaimed or directly re-used' mercury for uses specifically approved under the Convention.[21] The third and final convention to note in this context of recycling being based around very particular waste streams is the 2009 Convention for the Safe and Environmentally Sound Recycling of Ships. This Convention, which covers the recycling of an entire class of transport system, recognises clearly that 'recycling of ships contributes to sustainable development and, as such, is the best option for ships that have reached the end of their operating life'.[22]

## 4. RECYCLING AND THE BASEL CONVENTION

The final, and arguably the most important, international regime which addresses recycling is the 1989 Basel Convention on the Control of Transboundary Movements of Hazardous Waste and Their Disposal. The

---

[20] See articles 4(a) and (b), 7 and 9(1)(a) of the Montreal Protocol. For the relevant decisions under the Montreal Protocol on recycling of ozone depleting substances, see IV/26; V/15; VI/19; XVII/11; XX/7; and XXI/2.

[21] See article 11(b) of the 2013 Convention.

[22] Preamble. 2009 International Convention for the Safe and Environmentally Sound Recycling of Ships. SR/CONF/45. 19 May 2009. 'Ship Recycling', according to article 2 of the Convention, means the activity of complete or partial dismantling of a ship at a ship recycling facility in order to recover components and materials for reprocessing and reuse, whilst taking care of hazardous and other materials, and includes associated operations such as storage and treatment of components and materials on site, but not their further processing or disposal in separate facilities.

Basel Convention was negotiated to be a regime about controlling, not prohibiting, the disposal of waste in other countries. It was not, primarily, about recycling. Nevertheless, it was envisaged that the Parties could trade in waste, hazardous or otherwise, if 'the wastes in question are required as a raw material for recycling or recovery industries in the State of import'.[23]

To achieve this goal, an elaborate system of consent to take the waste of other nations was agreed. A key part of this consent being legitimate was if the importing country could suitably deal with the material that was exchanged. To assist in this area, thirteen identified operations which may lead to resource recovery, recycling, reclamation, direct reuse or alternative uses were identified via separate 'R' codes. The R codes encompass all such operations with respect to materials identified as hazardous under the Convention, which would otherwise be going for disposal (with different associated standards). The R codes include use as a fuel (other than in direct incineration) or other means to generate energy, the recycling/reclamation of metals and metal compounds as well as reclaiming, inter alia, solvents, oils, inorganic materials, acids, components for catalysts, and land treatment resulting in benefit to agriculture or ecological improvement. These operations are often accompanied by technical guidelines on how to achieve environmentally sound recycling practices within that particular resource type. For example, guidance exists on the environmentally sound recycling/reclamation of metals,[24] lead-acid batteries,[25] tyres[26] and electronic wastes.[27]

Despite such processes, widespread dissatisfaction arose with the Basel regime and how it was dealing with recycling soon after it was

---

[23] Article 4(9)(b).

[24] Basel Secretariat (2004). *Technical Guidelines on the Environmentally Sound Recycling/Reclamation of Metals and Metal Compounds.* (Basel Secretariat, Basel).

[25] Basel Secretariat (2003). *Technical Guidelines for the Environmentally Sound Management of Waste Lead Acid Batteries.* (Basel Secretariat, Basel).

[26] Basel Secretariat (1999). *Technical Guidelines on the Identification and Management of Used Tyres.* (Basel Secretariat, Basel). 3–4. Note, the guidelines were updated in 2011. See Appendices I and II in particular.

[27] Basel Convention (2012). *Technical Guidelines on Transboundary Movements of Electronic and Electrical Waste, In Particular, the Distinction Between Waste and Non-Waste*. (Basel Secretariat, Basel). Note, in 2013, it was decided to update these, with a particular regard to the distinction between waste and non-waste. See Decision BC-11/4 (2013). Technical Guidelines on Transboundary Movements of Electronic and Electrical Waste, In Particular, Regarding the Distinction Between Waste and Non-Waste under the Basel Convention.

established. This was because, in the words of Agenda 21, 'part of the international movement of hazardous wastes is being carried out in contravention of existing national legislation and international instruments to the detriment of the environment and public health of all countries, particularly developing countries'.[28] Although there was concern about the sending of waste for final disposal, part of the caution was about the growing trend of sending hazardous waste for recycling. Accordingly, Agenda 21 called upon the Parties to the Basel Convention, inter alia:

> To promote the development of control procedures for the transboundary movement of hazardous wastes destined for recovery operations under the Basel Convention that encourage environmentally and economically sound recycling options ... [via the] ... development of clear criteria and guidelines ... for environmentally and economically sound operations in resource recovery, recycling, reclamation, direct use or alternative uses and for determination of acceptable recovery practices, including recovery levels where feasible and appropriate, with a view to preventing abuses and false presentation.[29]

The concerns recorded in Agenda 21 quickly entered into the high level debates between the Parties at the Basel Convention. The initial concern for the Parties was how waste was being reclassified from final disposal (with higher standards – in terms of restrictions – for trade) to recovery (with lower standards for trade).[30] After significant debate, the majority of the Parties called, inter alia, for 'the phasing out by the end of 1997, all transboundary movements of hazardous wastes which are destined for recycling or recovery operations from OECD to non-OECD States'.[31]

---

[28] Agenda 21 (1992). UNCED. Doc A/CONF.151/4. Chapter 20, paragraph 4.

[29] Agenda 21 (1992). UNCED. Doc A/CONF.151/4. Chapter 20, paragraphs 33 and 34.

[30] For the initial concerns at the meetings, see Decision I/16. (1992). Transboundary Movement of Hazardous Wastes Destined for Recovery Operations. Decision I/22. (1992). Recovery Transfers and Environmentally Sound Processes. For commentary, see Clapp, J. (2001). *Toxic Exports*. (Cornell UP, NYC). 58–60.

[31] Decision II/12. (1994). Final Recovery Operations. Note also Decision II/14. (1994). Transboundary Movements of Hazardous Wastes Destined for Recovery Operations; Anon. (1994). 'Export of Hazardous Wastes'. *Environmental Policy and the Law*. 24(4): 147–8; MacKenzie, D. (1994). 'Europe Reluctant to Curb Toxic Waste Trade'. *New Scientist*. 2 Apr. 6.

A minority of developed countries objected to this due to uncertainties about what was hazardous (and should not be traded for recycling), and what was of potential economic value (and should be classified as safe to trade for recycling purposes), and the assumption that the principle of prior informed consent should allow states to make the decision about what they wish to import, or not, for themselves. Or, as the OECD suggested:

> Recovery of valuable materials and energy from wastes is an integral part of the international economic system and that well established international markets exist for the collection and processing of such materials within member countries.[32]

This minority of countries was also concerned about whether such a monumental decision that would fundamentally change the Convention could be made by resolution – as opposed to a formal amendment to the Convention.[33]

Following through, in 1995 the majority of the Parties seeking to prohibit the trade in hazardous waste for recycling purposes proposed a formal amendment. The proposed Article 4A,[34] known as the 'Ban Amendment', provided for the prohibition of exports of all hazardous

---

[32] Preamble. Decision C(92)39. For the subsequent iterations of this view, see the OECD Council Decision Concerning the Control of Transboundary Movements of Wastes Destined for Recovery Operations. C(2001)107/FINAL, as amended by C(2004)20; C(2005)141; and C(2008)156.

[33] MacKenzie, D. (1992). 'Business as Usual for Traders in Toxic Waste'. *New Scientist*. 2 Dec. 9; MacKenzie, D. (1993). 'Europe's Green Channel for Toxic Waste'. *New Scientist*. 13 Mar. 13–14; Anon. (1995). 'Recycling of Hazardous Waste'. *Environmental Policy and the Law*. 25(4); 184–5; Edwards, R. (1995). 'Leaks Expose Plan to Sabotage Waste Treaty'. *New Scientist*. 18 Feb. 3–4. For the NGO support for the ban, see IUCN. (1994). *Resolution 31: International Trade in Toxic Waste*. (19th General Assembly, Buenos Aires).

[34] 'Each Party listed in Annex VII shall prohibit all transboundary movements of hazardous wastes which are destined for operations according to Annex IV A to States not listed in Annex VII. Each Party listed in Annex VII shall phase out by 31 December 1997, and prohibit as of that date, all transboundary movements of hazardous wastes under Article 1(i)(a) of the Convention which are destined for operations according to Annex IV B to States not listed in Annex VII. Such transboundary movement shall not be prohibited unless the wastes are characterised as hazardous under the Convention.' Decision III/1. (1995). Amendment to the Basel Convention. See also Decision III/14. (1995). Transboundary Movements of Hazardous Wastes Destined for Recovery Operations.

wastes covered by the Convention that were intended for reuse, recycling and recovery from – in essence – developed countries to developing countries.

The problem was that, despite having a majority of Parties voting at that point to secure a resolution, they did not have the requisite numbers to formally amend the Convention.[35] Coupled with formal reservations about the proposed amendment due to the division of the countries, question marks over what was hazardous and whether countries should have the option of recycling removed from them, the situation turned into one of extreme confusion.

Although further work was set down to attempt to resolve the difficulties, all efforts trying to implement a ban on stopping all hazardous waste being traded for recycling stalled, and then stopped.[36] After several further meetings without agreement on this topic, the President of the meeting in 2009 issued a statement on the possible way forward, and affirmation, for the Ban Amendment. In this statement, the President called on the Parties to create enabling conditions, through amongst other measures, country-led initiatives conducive to the attainment of the objectives of the Amendment. Based on this, Indonesia and Switzerland announced their readiness to organise a 'Country-Led Initiative'. The results of these efforts formed the basis of the decision adopted in 2011.[37]

Section A of the 2011 decision addressed, without prejudice, the entry into force of the Ban Amendment and agreed an interpretation of article

---

[35] Any party may propose amendments (in accordance with a set format) to the Convention (article 17). If consensus is not possible, the amendment can be adopted by a three-fourths majority vote of the Parties present and voting at a conference of the Parties, or two-thirds for the Protocol (article 17.3). Amendments adopted 'shall enter into force between Parties having accepted them … or acceptance by at least three-fourths of the Parties who accepted them or by at least two thirds of the Parties to the protocol concerned who accepted them' (article 17.5). This issue and the words 'who accepted them' have been the subject of intense debate with regard to Decision III/1 over how best an amendment (and whether it should be by consensus, by entrenched majority, and if so – by whom) should be brought into force. See Decisions VIII/30 (2006) and IX/25 (2008) addressing the interpretation of paragraph 5 of article 17 of the Basel Convention.

[36] For attempts to keep the process going, see Decisions relating to Annex VII, see IV/8 (1998); V/4 (1999); VI/34 (1998); and VII/24 (2004). For commentary, see Clapp, J. (2001). *Toxic Exports*. (Cornell UP, NYC). 77.

[37] Decision BC-10/3 (2011). Indonesian-Swiss Country-led Initiative to Improve the Effectiveness of the Basel Convention.

17(5) of the Basel Convention on amendments to the Convention.[38] The rest of this wide-ranging decision called for, inter alia, the development of more guidelines around the environmentally sound management of hazardous wastes, capacity building, and greater legal clarity on the definitions of, amongst others, waste/non-waste, reuse, second-hand and used goods. This approach was followed up at the 2013 meeting, which passed a further decision on the matter.[39] This decision, whilst encouraging the Parties to ratify the Amendment, kept up the substantive work agreed to earlier, in terms of adopting the framework for the environmentally sound management of hazardous wastes and other wastes[40] and the creation of an expert working group[41] to work on related matters. Finally, working on the issue of legal clarity, a further inter-sessional group was created to work in this area, with a particular focus on the distinctions between waste and non-waste, specific problems like end-of-life computers, and the adequacy of existing recommendations and guidance from the Basel Convention on this work.

As the policy makers and lawyers have argued for the last 20 years over whether a ban for the purposes of recycling hazardous waste was in existence or not, the practice has developed whereby the answer to the question has been determined by the waste in question. This approach has followed the OECD position that a regulated trade in waste destined for recycling or recovery was entirely consistent with the Basel Convention.[42]

---

[38] Specifically, the Parties agreed 'without prejudice to any other multilateral environmental agreement, that the meaning of paragraph 5 of Article 17 of the Basel Convention should be interpreted to mean that the acceptance of three-fourths of those Parties that were Parties at the time of the adoption of the amendment is required for the entry into force of such amendment, noting that such an interpretation of paragraph 5 of Article 17 does not compel any party to ratify the Ban Amendment'.

[39] Decision BC-11/1 (2013). Follow-up to the Indonesian-Swiss Country-led Initiative to Improve the Effectiveness of the Basel Convention.

[40] This part of BC-11 came with a detailed Annex I, from Parties to Regional Centres, to begin substantive work on the environmentally sound management of wastes, at multiple levels.

[41] This part of BC-11 came with a detailed Annex II, which called for this group to work on, inter alia, guidelines, information collection and capacity building for environmentally sound management of hazardous wastes.

[42] Decision-Recommendation C(90)178/FINAL on the Reduction of Transfrontier Movements of Wastes (31 Jan. 1991). For commentary, see Krueger, J. (1999). *International Trade and the Basel Convention.* (Earthscan, London). 25–6.

The OECD position was, and is, based on a system of Green and Amber lists for wastes which could be recycled in other countries. Waste which required minimal procedural requirements (due to being a lesser risk) was placed on the Green list, whereas waste requiring greater levels of regulation was placed on the Amber list.[43] Thus, as the Europeans came to conclude, particular exports of waste for recycling or recovery to non-OECD countries was permissible if the correct procedures were followed and the waste was on an approved list.[44]

The Green list (which related, primarily, bar a few adjustments, to Annex IX in the Basel Convention) covers non-hazardous waste such as uncontaminated metals, textiles, paper, glass and some plastics. In the European context, the Green list contained, inter alia, metal-bearing wastes arising from melting, smelting and refining of metals (including electrical assemblies/scrap consisting only of metals or alloys), glass waste and solid plastic wastes. The procedural requirements of trade for recycling on the Green list are much less onerous than when dealing with the more contaminated forms of waste on the Amber list. The wastes on the Green list are deemed to pose negligible risks for human health and the environment. Accordingly, they are not subject to specific controls – although they should still be dealt with in an environmentally sound manner according to national and/or regional laws, regulations and practices.[45]

The Amber list (which relates, bar a few adjustments, to Basel Annexes II and VIII) covers wastes which, although not hazardous in themselves, still require careful handling. Trade can only take place upon prior written notification, consent and with tracking of the waste – as with all of the standard requirements of the Basel Convention. For the Europeans, to help ensure certainty in this area, the European Commission has passed a series of regulations clearly stating which wastes can go to which countries.[46]

---

[43] Note, originally there were three levels, with Red – clearly hazardous in which there should be no trade being the last addition. Subsequently, failure to qualify for either the Green or Amber list has meant, by deduction, that the substance is too hazardous to trade, such as wastes involving PCBs, asbestos, cyanide, and so on.

[44] Article 36. European Regulation 1013/2006 (14 June 2006) on the Shipments of Waste.

[45] See Chapter II, section C, and Annex III of European Regulation 1013/2006 (14 June 2006) on the Shipments of Waste.

[46] See Chapter II, section D, and Annex III of European Regulation 1013/2006 (14 June 2006) on the Shipments of Waste.

## 5. ECONOMIC AND ENVIRONMENTAL CONTEXTS

The global figure for the rate of recycling from municipal waste is 29 percent. Although an impressive overall figure, this generic number hides the differences between developed and developing countries. In the developed world, recycling is mainly carried out through high tech systems of processing, highly regulated regimes and is buttressed with strong social support as many citizens consider recycling almost a civic duty. When other incentives are placed to encourage recycling, such as price incentives to make dumping unattractive, ease of recycling, public education, quality control in recycled products and security of supply for recycled materials, the success rates can be relatively high. For example, in New Zealand, 94 percent of households claim to be involved in some forms of recycling. In the United States, an estimated 34 percent of all municipal waste is recycled (although in some cities, like San Francisco, the figure is closer to 70 percent). In Britain, the recycling sector has grown by more than 50 percent since 2004, amounting to an extra 48 million tons of material recovered, generating over £11 billion in 2011. Even the informal sector in Britain, such as car boot sales, is estimated to be effective in diverting some 50,000 to 60,000 tons of material away from final disposal, per year. Such high rates of recycling and reuse are also generating a considerable amount of formal employment, with at least 1 million people working in this sector in the United States.[47]

In the developing world, recycling is primarily done through informal, less regulated and more labour intensive settings. Thus, whilst households may have small interest in recycling, perhaps up to 2 million very lowly paid people scour all places of waste, obtaining anything of economic value that can be sold. Despite being paid very small sums for their finds, they manage to divert millions of tons of waste per year. For

---

[47] Gregson, N. (2013). 'Moving up the Waste Hierarchy: Car Boot Sales, Reuse and Exchange'. *Resources, Conservation and Recycling.* 77: 254–67; Seldman, N. (2012). 'Recycling Stimulates Economic Development'. *Biocycle.* 53(7): 22–5; World Bank (2012). *What a Waste: A Global Review of Solid Waste Management.* (World Bank, Urban Development Series, No. 15, Washington). viii, 5; Llerena, D. (2011). 'Green Consumer Behaviour: An Experimental Analysis of Willingness to Pay for Remanufactured Products'. *Business Strategy and the Environment.* 20(6): 178–92; Neo, H. (2010). 'The Potential of Large Scale Urban Waste Recycling'. *Society and Natural Resources.* 23: 872–87; Bohara, A. (2007). 'The Effect of Experience and Quantity-Based Pricing on the Valuation of a Curbside Recycling Program'. *Ecological Economics.* 64: 433–43; Meneses, G. (2006). 'How to Teach Recycling at an Advanced Phase of Diffusion'. *The Journal of Environmental Education.* 37(4): 19–32.

example, in the city of Delhi in India over 100,000 rag-pickers achieve recycling rates at waste disposal sites of between 50 to 90 percent for products which have a clear economic value. Likewise, ship-breaking employs over 1 million people worldwide, including 200,000 people in Bangladesh and 500,000 in India, each earning only a few dollars per day. In China, there are over 60,000 small scale family orientated workshops devoted to recycling plastic. The money these workers generate, although small by the standards in developed countries, especially for the risks involved (and the lack of comparable safety standards as found in the developed world), is very important to them and can be leveraged to achieve admirable personal and social goals which are important to those who undertake this work. This same principle – of people wanting to work in dangerous jobs because they pay better than alternative options – also applies to most of those who are most commonly illiterate and uneducated and work in the recycling sector in other parts of the developing world.[48]

Despite the strong support for recycling in both the developing and developed world, there are two distinct anomalies in this area. First, recycling can be seen as something of a distraction, as it does not address the greater problem, which is the exponential rise in waste. For example, whilst the volume of recycled consumer waste continues to rise (in the United States it rose from 5.6 to 65 million tons per year between 1960 and 2010), the mountain of overall waste continues to grow in the United States (from 88 million tons to 250 million).[49] Underlying part of this growth are considerations such as that the ability to recycle may actually encourage, not discourage, consumption in a number of instances. This impact, from consumers willing to consume more recycled products than they would if they were virgin, through to manufacturers of recycled

[48] Anon. (2014). 'Mucking It Up'. *Economist*. 5 Apr. 28; Rockson, G. (2013). 'Activities of Scavengers and Itinerant Buyers in Ghana'. *Habitat International*. 39: 148–55; Anon. (2013). 'Mucking In'. *Economist*. 2 Nov. 59; Oguntoyinbo, Y. (2012). 'Informal Waste Management Systems in Nigeria and Barriers'. *Public Health*. 126: 441–7; Paul, J. (2012). 'The Informal Sector in Municipal Solid Waste Management in the Philippines'. *Waste Management*. 32: 2018–28; UN Human Settlements Programme (2011). *Solid Waste Management in the World's Cities*. (UNSP, NYC). xxii, 127; Editor (2010). 'Waste, Poverty and Recycling'. *Waste Management*. 30: 171–3; Hayami, Y. (2006). Waste Pickers and Collectors in Delhi'. *Journal of Development Studies*. 42(1): 41–69; Wilson, D. (2006). 'The Role of Informal Sector Recycling in Waste Management in Developing Countries'. *Habitat International*. 30: 797–808.

[49] Minter, A. (2013). *Junkyard Planet*. (Bloomsbury, NYC). 12–13.

products seeking to accelerate demand for their products, does not address the base problems of over-consumption.[50]

The second anomaly is that there is a very clear difference between environmental- and economic-based recycling. Environmental-based recycling is where an individual, community or nation, because they do not wish an item to be subject to final disposal, chooses to pay to have a product changed into something else. Environmental-based recycling is relatively modern, with a lineage of only decades, and is predominantly the interest of the developed world. With environmental-based recycling, the economic costs are secondary considerations.

Economic-based recycling is where products are changed into something else, because of their economic value. In the 21st century, the global market in post-consumer (not even industrial or commercial) waste for metal, plastic and paper is put at, at least, 30 billion USD per year with supply and demand being subject to international competition. Although the economic size of this market has never been seen before, the market is not at all new. Individuals, communities and countries have been involved in recycling for economic benefits for hundreds of years. The specific economic benefits to be reaped are, potentially, cost avoidance (for disposal fees, reduced energy costs), and potentially increased sales and productivity in terms of market niches. Additional justification for recycling includes corporate image for commercial enterprises. However, the biggest determining factor on whether a profit will be made will be the market for recycled goods, and the cost of non-recycled alternatives. The bottom line is that in this area, unless there is a strong political and/or legal mandate to recycle something without recourse to economic considerations, the recycling efforts will fail. Money powers the process as surely as the sun's energy powers the winds. Absent the money, recycling – more often than not – lies becalmed. It is for this reason that whilst recycling rates are increasing in some areas, in many others, they are volatile, or declining.[51]

---

[50] Baril, A. (2014). 'Moving from Recycling to Waste Prevention'. *Waste Management and Research.* 32(9): 3–18.

[51] Kaffine, D. (2014). 'Scrap Prices, Waste and Recycling Policy'. *Land Economics.* 90(1): 303–24; Wood, R. (2013). 'Going for Zero: A Comparative Critical Analysis of Zero Waste Events'. *Australasian Journal of Environmental Management.* 10(1): 46–55; Kenwood, N. (2013). 'UK Seeks a Zero Waste Economy'. *Materials Resource Society Bulletin.* 38(2): 15–17; Jacobson, J. (2011). 'Education Important for Recycling Efforts'. *Beverage Industry.* 102: 42–4; Bohm, B. (2010). 'The Costs of Municipal Waste and Recycling Programs'. *Resources, Conservation and Recycling.* 54: 864–71; Editor (2009).

## 6. WHERE THE MARKET IS TEPID

There are four good examples of where the market for recycling is tepid and in need of encouragement. These areas, despite long-standing practices and environmental benefits, include the recycling of paper, glass, rubber and biowaste.

The recycling of biowaste/old food is as old as the first time a human threw some scrap food to an animal. From the 15th century, allowing animals such as goats and especially pigs (75 pigs can eat one ton of waste per day) to wander freely in cities to consume whatever was edible was common, up to the middle of the 19th century. Glass blowers have known for over 1,000 years that a clean supply of broken glass and the appropriate additives can be used in place of sand to make new glass. This recycling industry grew quickly through the industrial revolution, as did paper recycling. This industry can be traced to the end of the 17th century, where linen and cotton rags constituted the most important raw materials for the paper industry. By the middle of the 19th century, 65 percent of the paper in the United States came from recycled materials, with the rag trade between Britain and the United States, at 1 million pounds (by weight) per year, being the dominant source.[52]

---

'Let's Junk the Throwaway Society'. *New Scientist.* 19 Sept. 5; Lavee, D. (2009). 'The Effect of Recycling Price Uncertainty on Municipal Waste Management Choices'. *Journal of Environmental Management.* 90: 3599–606; Brahic, C. (2009). 'Follow that Trash'. *New Scientist.* 19 Sept. 36; Suttibak, S. (2008). 'Assessment of Factors Influencing the Performance of Solid Waste Recycling Programs'. *Resources, Conservation and Recycling.* 53: 45–56; Martin, H. (2006). 'Social, Cultural and Structural Influences on Household Waste Recycling'. *Resources, Conservation and Recycling* 48: 357–95; Loughlin, D. (2006). 'Policies for Strengthening Markets for Recyclables: A Worldwide Perspective'. *Critical Reviews in Environmental Science and Technology.* 36: 287–326; OECD (2005). *Improving Recycling Markets.* (OECD, Paris). ENV/EPOC/WGWPR (2005)3/FINAL; Porter, R. (2002). *The Economics of Waste.* (Resources for the Future, Washington). 162–7; Frosch, R. (1995). 'The Industrial Ecology of the 21st Century'. *Scientific American.* September. 144–8; Boerner, C. (1994). 'The Folly of Demand Side Recycling'. *Environment.* 36(1): 7–15; Oliver, J. (1975).'The Value of Recycling: A Framework for Analysis'. *Industrial Marketing Management.* 4: 133–41.

[52] MacBride, S. (2012). *Recycling Reconsidered.* (MIT Press, Boston). 32–3; Franchetti, M. (2009). *Solid Waste Analysis and Minimisation.* (McGraw Hill, NYC). 7, 25, 56–8; Medina, M. (2001). 'Scavenging in America: Back to the Future?'. *Resources, Conservation and Recycling.* 31: 229–40; Cooper, T. (2006). 'Rags, Bones and Recycling Bins'. *History Today* 56(2): 17–18; Rogers, H. (2005). *Gone Tomorrow: The Hidden Life of Garbage.* (Norton, NYC). 172–5.

Of all of these four recycling industries, there are strong environmental benefits in recycling. In terms of landfill, old paper is a massive waste stream. For example, in the United States, this stream is at least 46 million tons of old newspapers, magazines, office papers and boxes per year. Globally, this stream makes up, on average, 17 percent of the total in any standard dump. Glass makes up on average 5 percent of landfill (it used to be higher, at 30 percent, but has been eclipsed by plastic). Rubber tyres are a particular plague for dumps, holding insects, vermin and easily catching fire. Below ground, they leach, are difficult to keep entombed, do not disintegrate and can take up disproportionate amounts of space. Finally, biowaste produces the lion's share of the waste stream. This can be as much as 64 percent of the household waste generated in developing countries, at up to 7 kilograms per person, per day. Conversely, in the high income countries, biowaste is only about 28 percent of the municipal waste stream. In Europe, about one third of all purchased food is later discarded (at over 116 million tons per year).[53]

In terms of environmental benefits based upon recycling old material, as opposed to production based upon virgin sources, old paper is particularly obvious. To make a paper product from new sources, as opposed to recycled, requires more energy (64 percent), water (50 percent), chemicals (20 times more chlorine) and makes more air pollution (74 percent). Finally, the question of where virgin paper is sourced from comes to the fore, because if it is from old growth or tropical forests, the environmental implications are even larger. Every ton of office paper recycled is believed to be the equivalent of 24 trees that can be left standing. The difficulty with this recycling is that this market is continually threatened by economically cheaper sources of virgin material, which undercuts the environmental benefits. The same problem exists for recycled glass and is likely to worsen as the difficulties and costs of recycling new glass streams, such as cathode ray tubes (CRTs)/ old style television projectors, make the economic costs of the recycled product even greater than virgin material. Similar problems exist also for the products made of recycled rubber and for composting initiatives, in

---

[53] Gentil, E. (2012). 'To Waste or Not to Waste: Food'. *Waste Management and Research*. 30(5): 455–7; Oelofse, S. (2013). 'Estimating the Magnitude of Food Waste Generated in South Africa'. *Waste Management and Research*. 31(1): 80–86; World Bank (2012). *What a Waste: A Global Review of Solid Waste Management.* (World Bank, Urban Development Series, No. 15, Washington). 19; Anon. (2010). 'Managing Organic Municipal Waste'. *Appropriate Technology*. 37(3): 62–7; Diaz, L. (2006). 'Managing the Organic Fraction of Municipal Solid Waste'. *BioCycle*. 47(10): 50–52.

that the economic costs of collecting and making the compost are often much greater than what it can be sold for.[54]

In all four areas, the recycling industries are either temperamental or kept active by government targets by which recycling rates must be achieved. For example, in Europe, 50 percent of all consumer glass, by weight, must be recycled by 2020. Targets on paper recycling are 60 percent from packaging and 50 percent from other types of consumer waste paper, by weight, by 2020. Tyres must be kept out of the waste stream and up to 65 percent of the biowastes must be directed away from

---

54 Ling, T. (2013). 'Management and Recycling of Waste Glass'. *Resources, Conservation and Recycling*. 70: 25–31; World Bank (2012). *What a Waste: A Global Review of Solid Waste Management.* (World Bank, Urban Development Series, No. 15, Washington). 19, 27; Blengini, G. (2012). 'Eco-Efficient Waste Glass Recycling'. *Waste Management*. 32: 1000–1008; European Paper Recovery Council (2012). *Recycled Paper: The European Declaration on Paper Recycling, 2010–2015*. (EPRC, Geneva). v–vii; MacBride, S. (2012). *Recycling Reconsidered.* (MIT Press, Boston). 23–5, 28–33, 41; Wang, L. (2012). 'A Life Cycle Assessment (LCA) Management Options for Waste Papers'. *Bioresource Technology*. 120: 89–98; Kellenberg, D. (2012). 'Trading Wastes'. *Journal of Environmental Economics and Management* 64: 68–87; Mueller, J. (2012). 'CRT Waste Glass Processing'. *Waste Management*. 32: 1560–65; Chao, C. (2011). 'Reduce Cost for Recycling Glass'. *Waste Management*. 31: 2414–21; Farrin, L. (2010). 'The Environmental Benefits From Reusing Clothing'. *The International Journal of Life Cycle Assessment*. 15(7): 726–36; Anon. (2010). 'Paper Wastage'. *New Scientist*. 24 Apr. 5; Franchetti, M. (2009). *Solid Waste Analysis and Minimisation*. (McGraw Hill, NYC). 76–7; Yamaguchi, M. (2009). 'The Blending Ratio of Waste Paper'. *Waste Management*. 29: 1491–4; Monte, M. (2009). 'Waste Management from Pulp and Paper Production'. *Waste Management*. 29: 293–312; Vellini, M. (2009). 'Glass Container Production and Recycling'. *Energy*. 34: 2137–43; Cherry, P. (2009). 'The Next Billion Computers'. *Journal of Cleaner Production*. 17: 1620–28; Ismail, Z. (2009). 'Recycling of Waste Glass'. *Waste Management*. 29: 655–9; Villanueva, A. (2007). 'Paper Waste: Recycling?'. *Waste Management*. 27: S29–S46; Anon. (2007). 'Saint and Sinner on Paper'. *New Scientist*. 21 July. 6; Claudio, L. (2007). 'The Environmental Impact of the Clothing Industry'. *Environmental Health Perspectives*. 115(9): A448–A454; Rogers, H. (2005). *Gone Tomorrow: The Hidden Life of Garbage*. (Norton, NYC). 134–5, 148–50; Schenk, N. (2004). 'Paper Recycling and Primary Pulp Requirements'. *Journal of Industrial Ecology* 8(3): 141–50; Hansen, K. (2004). 'Controversies Around the Second-Hand Clothing Trade'. *Anthropology Today*. 20(4): 3–10; Samakovlis, E. (2004). 'Revaluing the Hierarchy of Paper Recycling'. *Energy Economics*. 26: 101–22; Baumgarthner, S. (2003). 'Price Ambivalence of Waste Paper in Germany'. *Ecological Economics*. 47: 183–95; Bystrom, S. (2000). 'Paper Recycling'. *Resources, Conservation and Recycling*. 28: 55–65.

landfills by 2016.[55] In all instances, these targets are supplemented by initiatives which seek to stimulate demand for recycled products. These range from collection to quality control, and some guaranteed end-uses such as a certain percentage of phone directories being made from recycled paper, or motorway surfaces from recycled tyres. Alternatively, demand may be created. For example, former President Bush sought to stimulate recycling by mandating the procurement of recycled goods in all federal agencies. This was similar to President Clinton's requirement that federal agencies support recycling programmes by using recycled products and services wherever possible.[56]

## 7. WHERE THE MARKET IS VORACIOUS

Unlike the above examples where the market for recycling is tepid and often in need of direct support to ensure that the economic benefits are made worthwhile, in at least two areas, those of metal and plastic, the market is voracious and largely self-driven due to the economic benefits of recycling. However, if poorly managed, these recycling practices, which have become strongly international, can cause large scale environmental and/or social damage. It is because of these risks that the international community has attempted to control the market.

---

[55] See articles 11 and 22 of Directive 2008/98 (19 Nov. 2008) on Waste and article 5 of Directive 1999/31/EC (26 Apr. 1999) on Landfill. The European standards are in article 6 of Directive 2004/12/EC (20 Dec. 1994) on Packaging and Packaging Waste. For the question on restrictions into landfill, see EPA (1999). *State Scrap Tire Programmes*. (EPA, Washington). 4. For the European example, see article 4(3)(a) of Directive 1999/31/EC (26 Apr. 1999) on Landfill – prohibiting the disposal of whole used tyres. In 2003, this ban was extended to shredded tyres. See also, World Bank (2012). *What a Waste: A Global Review of Solid Waste Management*. (World Bank, Urban Development Series, No. 15, Washington). 5.

[56] For Bush and Clinton, see Executive Orders 12780 and 12873. Also, EPA (2012). *Border 2012: US-Mexico, Border Scrap Tyre Inventory*. (EPA, Washington). 5; EPA (2011). *Municipal Solid Waste Generation, Recycling, and Disposal in the United States: Facts and Figures for 2010*. (EPA, Washington, EPA-530-F-11-005). 3–4; WTO (2007). *Measures Affecting Imports of Retreated Tyres*. (WTO, Geneva) WT/DS332/AB/R. 3 December 2007; Porter, R. (2002). *The Economics of Waste*. (Resources for the Future, Washington). 123; EPA (1991). *Markets for Scrap Tyres*. (EPA, Washington). 3–4; Strasser, S. (1999). *Waste and Want: A Social History of Trash*. (Holt, NYC). 102–3, 106–8, 240–41; Mestel, R. (1995). 'Old Tyres Keep Fairways Green'. *New Scientist*. 22 Apr. 6; Burke, M. (1993). 'A Cool Way to Treat Worn Out Tyres'. *New Scientist*. 26 June. 19.

The national and international markets for the recycling of plastic are growing at great speed. Most consumers are broadly aware of the recycling possibilities in this area, coming into constant contact with the numeric codes (seven in all, corresponding to six different polymers, and a seventh catch-all category) stamped on each plastic item, with three chasing arrows. The environmental and economic benefits of recycling plastic are strong. Specifically, compared to plastics made from virgin materials, recycled plastic materials generally require about three times less energy, eight times less water and produce 50 percent less carbon dioxide. Recycling also keeps plastic out of the ocean and landfill (of which it is currently about 10 percent of the standard municipal waste stream).

Success in this area is evident with two plastics in particular, namely, polyethylene terephthalate/PET (number 1 code) and rigid containers, usually high-density polyethylene/HDPE (number 2 code). In some European countries, the recycling rate of PET bottles is close to 80 percent, whilst for the United States some 29.2 percent of PET bottles are being recycled, as are 27.5 percent of natural white HDPE bottles. The recycling rates of PET can be expected to rise as dominant multinationals such as Coca-Cola have pledged themselves to move from their current levels of recycled PET content to eventually 100 percent. In addition to being recast as new PET bottles, PET is also recycled into fibre, clothing, thermal insulation and strapping. Recycled HDPE often ends up as furniture, buckets, bins, drums, and drainage pipes. These successes are due to the fact that the raw price of virgin resin (which is directly linked to the price of oil and/or natural gas) for these plastics is typically higher than that of the recycled product. The price of the recycled product can be kept low if the recycled plastic is not contaminated with other types of plastic, is largely uniform and pre-sorted beforehand. It is for this reason that countries like China (which takes 70 percent of this global trade) have banned the import of whole plastic bottles, only allowing that which has been washed, pre-shredded and put through initial sorting procedures in the countries of origin to enter Chinese ports, via licensed dealers.[57]

---

[57] Duraiappah, A. (2011). 'Issues in Production, Recycling and International Trade of Plastic'. *Environment and Development Economics.* 7(1): 47–74; Freinkel, S. (2011). *Plastic* (Houghton, NYC). 73–80, 175, 183, 201–2; Intini, F. (2011). 'Insulation Made of Recycled Post-Consumer PET Bottles'. *International Journal of Life Cycle Assessment.* 16: 306–15; Frigione, M. (2010). 'Recycling of PET Bottles'. *Waste Management.* 30: 1101–6; Welle, J. (2011). 'Twenty Years of PET Bottle to Bottle Recycling'. *Resources, Conservation and*

Although PET and HDPE are making recycling progress, the other plastics – about 60 percent of the total plastic production – are only making slow progress in recycling, and although often technologically possible, are currently less enticing in economic terms. Accordingly, as the market alone will not solve this problem, some regions, countries or individual municipalities have set targets for the recycling of plastic. These objectives are supported by international goals which recognise the value of recycling plastic waste.[58] This is supported by technical guidelines from the Basel Convention, in an attempt to make this as environmentally and socially safe as possible.[59] In addition, regionally, Europe set a broad target of 2008 for recycling 22.5 percent of all plastics associated with packaging and 50 percent for all plastics associated with post-consumer waste in general.[60]

The second national and international area which is very strong in recycling terms involves metal. Metal recycling is as old as metal-craft.

---

*Recycling*. 55: 865–75; Teschler, L. (2009). 'World's Biggest Plastic-Bottle Recycling Plant'. *Machine Design*. 81: 52; Reilly, M. (2007). 'Recycled Plastic to Get Clean Bill of Health'. *New Scientist*. 12 May. 28–9; Pellow, D. (2007). *Resisting Global Toxics*. (MIT Press, Cambridge). 134, 136, 138–40, 142; Park, C. (2007). 'PVC Removal from Mixed Plastics by Separation'. *Journal of Hazardous Materials*. 144: 470–76; Tachwali, Y. (2007). 'Automatic System for Plastic Bottles Recycling'. *Resources, Conservation and Recycling*. 52: 266–85; Awaja, F. (2005). 'Recycling of PET'. *European Polymer Journal*. 41: 1453–77; Ambrose, C. (2002). 'Diversion from Landfill: Quality Products from Valuable Plastics'. *Resources, Conservation and Recycling*. 36: 309–18; Howell, J. (1992). 'A Ten Year Review of Plastics Recycling'. *Journal of Hazardous Materials*. 29: 143–64.

58 RIO+20. A/CONF.216/XX. Paragraph 218.

59 Basel Convention (2002). *Technical Guidelines for the Identification and Environmentally Sound Management of Plastic Wastes and for their Disposal*. (Basel Convention Secretariat, Basel). 12–14, 20–22, 24–8.

60 See article 6 of Directive 2004/12/EC (20 Dec. 1994) on Packaging and Packaging Waste. Note, plastics and glass with heavy metal concentrations have a derogation. See also article 11 of Directive 2008/98/EC (19 Nov. 2008) on Waste. With particular rules on identification, see Directive 97/129/EC (28 Feb. 1996) on Packaging. For derogation for plastic crates and pallets with heavy metal concentrations, Directive 2009/292/EC (20 Dec. 2009) on Packaging and Packaging Waste. For the facts and figures on plastic recycling, see World Bank (2012). *What a Waste: A Global Review of Solid Waste Management*. (World Bank, Urban Development Series, No. 15, Washington). 19; Rogers, H. (2005). *Gone Tomorrow: The Hidden Life of Garbage*. (Norton, NYC). 180–81; Zandonella, C. (1999). 'Plastic Recycling Toughens up Its Act'. *New Scientist*. 3 Apr. 26; Coghlan, A. (1992). 'Recycling Benefits Are in the Bag'. *New Scientist*. 7 Nov. 19.

Coins and jewellery made of gold and silver were melted down in ancient times to make new coins, albeit with images of the new sovereign. Other metals, too, were reused in preindustrial societies. Mining of new ore was dangerous and expensive, not always possible and generally much more difficult that the alternative of reusing what already existed. Metal objects have been too valuable to simply discard. Rather, when of no further use in their original incarnation, they have been traded, locally, nationally and internationally (since the 16th century) whereby they have been cleaned, melted and then reused by blacksmiths and other craftspeople. Thus, by the end of the last decade of the 19th century as railways were streaking across the United States, over 380,000 tons of scrap iron and steel was being imported into America. The justification for the recycling then, as it is now, was purely economic. It has also become evident since the end of the 20th century that, despite risks to the labour force in general metal recycling, there are also strong environmental benefits in recycling metal, as depending on type, recycling uses much less energy and creates less pollution than the manufacture of new metals from raw sources. For example, recycling 1 ton of aluminium cans conserves the equivalent of 36 barrels of oil. Or, when looking at a country like China, the recycling of metal in the first decade of the 21st century (not including iron and steel) saved them 110 million tons of coal and the need to excavate 9 billion tons of ore. On the aluminium equation alone, the pursuit of the recycled option prevented 552 million tons of carbon dioxide being released.[61]

The result, due in large part to the economic benefits of recycling, is that although metal only makes up a small part of the municipal waste stream (4 percent by global average) it is the most actively sought after discarded product for recycling. More than 400 million tons of post-consumer scrap metal are collected, globally, each year. When old vehicles (14 million per year in America alone), vessels (about 800, globally, per year), old planes (about 3,000 per year, again globally) and wastes from industrial usages are added, a sum of some 50 billion tons per year is produced. Recycling of aluminium often exceeds 50 percent (with less than 15 percent coming from consumer waste in the form of beer and soda cans). Over 80 percent of modern vehicles end up being stripped, compressed, melted down and recast (rather than being dumped,

---

[61] Minter, A. (2013). *Junkyard Planet*. (Bloomsbury, NYC). 7.

of which if they were, in Europe alone, a further 9 million tons of waste would have to be dealt with).[62]

Likewise, for lead-acid batteries that are commonly used in motor vehicles, the recycling rates are 96 percent in the United States and 85 percent in Europe. Aside from the leachate problems these were creating in landfills, recycled lead-acid batteries are about 25 percent cheaper to make than virgin ones. The result is most modern lead-acid batteries contain 60 to 80 percent recycled lead.[63]

A very similar trend has occurred with electronic waste. Until the year 2000 it was common to dump electronic waste. This dumping was often responsible for up to 70 percent of all heavy metals going into landfills. This type of practice has largely abated as entrepreneurs have become aware of the economic benefits to be found in recycling the gold, silver, palladium, iron, copper, indium, gallium and other rare earth metals that reside in the circuit boards and associated parts of electronic waste. High end recycling methods of furnaces and solvents can recover up to 95 percent of these particularly valuable metals. Estimates suggest that one ton of scrap from discarded personal computers contains more gold than can be produced from 16 tons of ore. Similarly, with mobile phones, of their 40 percent content which is metal (40 percent is plastic, and 20 percent ceramics) valuable recycling can be achieved. Specifically, of every 1 million phones recycled, an estimated 35,274 pounds of copper,

---

[62] For commentary, see Anon. (2012). 'Hard to Break Up'. *Economist*. 27 Oct. 26; Xiang, J. (2011). 'Vehicle Remanufacturing in China'. *Journal of Cleaner Production*. 19: 680–86; Gerrard, J. (2007). 'European End-of-Life Vehicle Legislation'. *Journal of Cleaner Production*. 15: 17–27; Wang, L. (2013). 'Policies and Perspective on End-of-Life Vehicles'. *Journal of Cleaner Production*. 7: 82–90; Che, J. (2011). 'End-of-Life Vehicle Recycling and International Cooperation'. *Journal of Environmental Sciences*. 23: 162–6.

[63] UNEP (2011). *Study on the Possible Effects on Human Health and the Environment in Asia and the Pacific of the Trade of Products Containing Lead, Cadmium and Mercury*. (UNEP, Nairobi). 12; Genaidy, G. (2008). 'Lead Recovery in Lead-Acid Battery Lifecycle'. *Science of the Total Environment*. 4(7). 7–12; Porter, R. (2002). *The Economics of Waste*. (Resources for the Future, Washington). 200; Anon. (2007). 'Toxic Down Under'. *New Scientist*. 7 July. 4. For an example of legal targets in this area in Europe, Directive 2006/66/EC (6 Sept. 2006) on Batteries and Accumulators stipulates that recycling 65 percent by average weight of lead-acid batteries and accumulators was the target. See Annex III, Part B, Recycling.

772 pounds of silver, 75 pounds of gold and 33 pounds of palladium can be recovered.[64]

All of these metals are economically and socially valuable, providing both money and raw materials to many countries. For example, shipbreaking provides 15 percent of India's steel and 80 percent of that for Bangladesh. Similarly for China, of the 5.6 million tons of copper produced in 2012, 2.75 million was made from scrap (and of that scrap, 70 percent was imported). Even the United States obtains up to 30 percent of its new steel from recycled sources.[65]

---

[64] Anon. (2013). 'A Cadmium Lining'. *Economist.* 26 Jan. 52; Leigh, N. (2012). 'Electronic Waste Recycling'. *Journal of Industrial Ecology.* 16(6): 940–50; UNEP (2011). *Where Are WEee in Africa? E-Waste Africa Programme.* (UNEP, Nairobi). 16–18; Pellow, D. (2007). *Resisting Global Toxics.* (MIT Press, Cambridge). 187; Graham-Rowe, D. (2006). 'Hazards in Our Electronic Gadgets'. *New Scientist.* 23 Sept. 26–7; Sinha, D. (2005). 'Electronic Waste Recycling'. *Environmental Impact Assessment Review.* 25: 492–504; Fischetti, M. (2009). 'Old Circuit Boards Hit Road'. *Scientific American Earth 3.0.* 19(2): 10; Douglas, E. (2008). 'Digging up the Dump'. *New Scientist.* 4 Oct. 34; Anon. (2008). 'China's Heavy Metal Overdose'. *New Scientist.* 5 Apr. 23; Widmer, R. (2005). 'Global Perspectives on E-Waste'. *Environmental Impact Assessment Review.* 25: 436–58.

[65] International Resource Panel. (2013). *Metal Recycling: Opportunities, Limits, Infrastructure.* (UNEP, Nairobi). 22–34; Reck, B. (2012). 'Challenges in Metal Recycling'. *Science.* 10 Aug. 337: 690–95; Townend, W. (2011). 'Metal Recycling, the Environment and Human Health'. *Waste Management Research.* 29: 675–80; Vincent, W. (2008). 'Scrap Metal Recycling'. *The Technology Teacher.* 67(5): 13–17; Jui, C. (2010). 'Recycling of Automotive Aluminum'. *Transactions of the Nonferrous Metals Society.* 20: 2057–65; Kindzierski, W. (2009). 'Automotive Wastes'. *Water Environment Research.* 81: 1545–68; Vincent, W. (2008). 'Scrap Metal Recycling'. *The Technology Teacher.* 67(5): 13–17; Douglas, E. (2008). 'Digging up the Dump'. *New Scientist.* 4 Oct. 34; Giannouli, M. (2007). 'Waste from Road Transport'. *Journal of Cleaner Production.* 15: 1169–82; Mikelis, N. (2007). 'A Statistical Overview of Ship Recycling'. *IMO News.* 4: 25–30; Rogers, H. (2005). *Gone Tomorrow: The Hidden Life of Garbage.* (Norton, NYC). 14, 204; Strasser, S. (1999). *Waste and Want: A Social History of Trash.* (Holt, NYC). 97–103, 153–5, 230, 235, 242–9; Porter, R. (2002). *The Economics of Waste.* (Resources for the Future, Washington). 184–90; Bellmann, K. (2000). 'Economic Issues in Recycling End-of-Life Vehicles'. *Technovation.* 20: 677–90; EPA (2011). *Municipal Solid Waste Generation, Recycling, and Disposal in the United States: Facts and Figures for 2010.* (EPA, Washington, EPA-530-F-11-005). 3–4; Anon. (1996). 'Recycling Old Refrigerators'. *Machine Design.* 68: 8; Steuteville, R. (1993). 'The Economics of Aluminum Recycling'. *BioCycle.* 34(1): 74; Rathje, W. (1992). *Rubbish: The Archeology of Garbage.* (Harper, NYC). 198–9; Carlsen, E. (1981). 'The

Due to such strong types of benefits, government intervention, especially by encouraging recycling via target setting, is not required. The exception to this is Europe, with recycling targets for scrap vehicles (95 percent by average weight, by class, by 2015)[66] in addition to targets of 50 percent for all metal wastes related to packaging and other forms of consumer waste.[67] Similarly, Europe has created recycling obligations which aim to collect up to 85 percent of most types of electronic waste by 2018.[68] Such target setting and obligatory recycling goals are also evident with small scale, consumer type batteries. Here, the economic benefits of recycling are not yet convincing, although it is expected that, as new technologies evolve, the recycling of the cadmium, mercury, lithium, manganese, nickel and zinc in these smaller batteries will become much more compelling. It is hoped that the same patterns will apply to the recycling of the (rapidly escalating) supply of batteries used in electric and/or hybrid motor vehicles.[69]

Due to the fact that the economics of recycling small consumer type batteries are not yet convincing, but the environmental benefits are, a number of governments and/or regions in the United States, Canada,

---

Recycling of Nonferrous Metals in the U.S Economy: 1959–1976'. *Conservation and Recycling*. 4: 1–8.

[66] See articles 5, 6 and 7(2)(b) of Directive 2000/53/EC (18 Sept. 2000) on End-of-Life Vehicles.

[67] See article 6 of Directive 2004/12/EC (20 Dec 1994) on Packaging and Packaging Waste. See also article 11 of Directive 2008/98/EC (19 Nov. 2008) on Waste.

[68] Articles 5, 6, 7, 8, 9 and 11 and Annex V. Directive 2012/19/EU (4 July 2012) on Waste Electrical and Electronic Equipment.

[69] Foster, M. (2014). 'Feasibility Assessment of Remanufacturing, Repurposing and Recycling of End of Vehicle Application Lithium Batteries'. *Journal of Industrial Engineering and Management*. 7(3): 79–92; Krekeler, M. (2012). 'Spent Alkaline Battery Recycling'. *Journal of Power Sources*. 203: 222–6; Belardi, G. (2011). 'Characterization of Spent Zinc-Carbon and Alkaline Batteries'. *Thermochimica Acta*. 526: 169–77; Kuo, Y. (2009). 'Reclaiming Spent Alkaline Batteries'. *Waste Management*. 29(7): 2132–9; Shin, S. (2009). 'Separation of Zinc from Spent Zinc-Carbon Batteries'. *Hydrometallurgy*. 96: 349–53; Paulino, J. (2008). 'Recovery of Valuable Elements from Spent Li-Batteries'. *Journal of Hazardous Materials*. 150(3): 843–9; Ferella, F. (2008). 'Process for the Recycling of Alkaline and Zinc-Carbon Spent Batteries'. *Journal of Power Sources*. 183(2): 805–11; Bertuol, F. (2006). 'Spent NiMH Batteries'. *Journal of Power Sources*. 160: 1465–70; Bernades, R. (2004). 'Recycling of Batteries'. *Journal of Power Sources*. 130: 291–8; Tenorio, J. (2002). 'Recovery of Ni-Based Alloys from Spent NiMH Batteries'. *Journal of Power Sources*. 108(1): 70–73.

Australia and Europe have set recycling targets for this waste stream. In the case of Europe, all member states have to maximise the separate collection of waste batteries and accumulators and to minimise the disposal of batteries and accumulators as municipal waste in order to achieve a high level of recycling. This must be done at no cost for the last purchaser of the working product. In terms of collection schemes, this meant that each member state had to enable end-users to discard waste portable batteries or accumulators at suitable collection points and all distributors were required to take back waste portable batteries or accumulators at no charge. The recycling targets are 75 percent by average weight (by 2016) of nickel-cadmium batteries and accumulators and 50 percent by average weight of other waste batteries and accumulators.[70]

Although the economic benefit of recycling metals has been evident for a long period and this market largely operates without governmental intervention, this industry can have impacts in both environmental and social terms, especially when expanded to a global level.[71] This has meant that the international community has attempted to influence how the negative environmental and social costs of recycling may be confronted through two separate conventions. The first, the Basel Convention, deals with the recycling of metals which may have hazardous implications, with particular regard to lead-acid batteries and electronic waste. The second deals specifically with the recycling of ships.

The influence of the Basel Convention in this area comes in three parts. The first deals with its Technical Guidelines on the Environmentally Sound Recycling/Reclamation of Metals.[72] These guidelines, which attempt to cover all of the possible forms of metal recycling that may be hazardous, draw out the particular risks for the dozen heavy metals listed in Annex I of the Convention. They then go on to provide guidance on the standards for what an environmentally sound reclamation facility

---

70 See articles 8 and 10 of Directive 2006/66/EC (6 Sept. 2006) on Batteries and Accumulators. See also Annex III.

71 Raun, L. (2013). 'Unanticipated Potential Cancer Risk near Metal Recycling Facilities'. *Environmental Impact Assessment Review*. 41: 70–77; Reck, B. (2012). 'Challenges in Metal Recycling'. *Science*. 337(6095): 960–70; Ayres, R. (1997). 'Metals Recycling: Economic and Environmental Implications'. *Resources, Conservation and Recycling*. 21(3): 145–73.

72 Basel Secretariat (2004). *Technical Guidelines on the Environmentally Sound Recycling/Reclamation of Metals and Metal Compounds*. (Basel Secretariat, Basel).

should encompass, including environmental and health considerations (ranging from how to deal with wastes and residue through to emergency situations).

The second example of the influence of the Basel Convention in this area of providing guidance to Parties interested in metal recycling relates specifically to lead-acid batteries. The international interest in this problem follows five decades of domestic experience in a number of countries that has shown that the breaking and recycling of these batteries – even after they were drained of acid – could be particularly dangerous. Due to such concerns, by the 1980s in the developed world, the practices of recycling the batteries had become entirely mechanised, with automated feeds within tightly sealed mills that were flushed continually with water. Human hands and lungs were kept away from the hazards. In the barely regulated informal sectors seeking to recycle lead-acid batteries in the developing world, such standards and/or appropriate technology are often not applied. Accordingly, spills of the acid electrolyte and associated lead dust contamination are common. Similarly, once the lead has been obtained, if it is refined poorly, dangerous vapour will be created. All of these have significant implications for human and environmental health.[73] It is because of these social and environmental risks that the Parties to the Basel Convention created its Technical Guidelines on Recycling Lead-Acid Batteries to help countries develop necessary standards, the most fundamental of all being 'the manual breaking of

[73] Anon. (2011). 'Activists Urge Tighter Waste Restriction on Used Lead Battery Exports'. *Inside EPA's Environmental Policy Alert*. 28:18; Were, F. (2010). 'Air and Blood Lead Levels in Lead Acid Battery Recycling and Manufacturing Plants in Kenya'. *Journal of Occupational and Environmental Hygiene*. 9: 340–44; Kreusch, M. (2007). 'Technological Improvements in Automotive Battery Recycling'. *Resources, Conservation and Recycling*. 52: 368–80; Haefliger, P. (2009). 'Mass Lead Intoxication from Informal Used Lead-Acid Battery Recycling in Dakar, Senegal'. *Environmental Health Perspectives*. 117(110): 1537; Simpson, J. (2003). 'Small Scale Lead Battery Refining Plants'. *Journal of Power Sources*. 30(4): 5–15; Clapp, J. (2001). *Toxic Exports*. (Cornell UP, NYC). 65; Warren, C. (2000). *Brush with Death: A Social History of Lead Poisoning*. (John Hopkins UP, London). 142–4; Elmer, J. (1996). 'The Basel Convention: Effect on the Asian Secondary Lead Industry'. *Journal of Power Sources*. 59(1): 1–7; Moyers, B. (1991). *Global Dumping Ground*. (Lutterworth, Cambridge). 54–8.

batteries should be avoided at all costs'.[74] How far these guidelines have improved the safety standards in this area is a matter of debate.[75]

The third example of the influence of the Basel Convention in this area of providing guidance to Parties interested in metal recycling relates specifically to electronic waste. This is a particularly interesting area, as it is the one type of metal recycling that was drawn out for words of caution at the 2012 Earth Summit where it was recognised that 'solid wastes, such as electronic waste, pose particular challenges, which should be addressed'.[76] The need to address this problem is because electronic waste often contains lead, mercury, cadmium, hexavalent chromium, polybrominated diphenylethers (PBDEs), polychlorinated and polybrominated dioxins (PCDD and PBDD) and brominated flame retardants. In addition to these risks, mobile phones also contain cadmium, lead, beryllium and antimony. These are not necessarily problems if their extraction is dealt with safely, by good, largely automated processes with robust environmental and labour standards. The risk is when this waste is disassembled, stripped and the valuable metals extracted in barely regulated, largely manual operations running through the informal sector, which lack the necessary infrastructure, capacity or resources. In these informal settings, circuit boards are often ripped out of old equipment without correct safety standards, heated over open fires or stripped via dipping them into barrels full of acid. Highly toxic residue wastes are then burnt in open pits. Such practices create negative impacts on the environment, workers and associated communities.[77]

---

[74] The Basel Convention Secretariat (2003). *Technical Guidelines for the Environmentally Sound Management of Waste Lead-Acid Batteries.* (BCS, Document No. 2003/9). Paragraph 41. Note, used vehicle batteries can be found under either List A, A1160 or List B, B1090, B1110 and B4030 of the Basel Convention.

[75] Stone, H. (2009). 'Effects of Amendments to the Basel Convention on Battery Recycling'. *Journal of Power Sources.* 78: 251–5; Hoffman, U. (2000). 'Battery Recycling in the Wake of Basel Convention Trade Restrictions'. *Journal of Power Sources.* 88: 115–23.

[76] RIO+20. A/CONF.216/XX. Paragraph 218.

[77] Luo, P. (2014). 'Health Risk Characterisation for Resident Inhalation Exposure in Typical E-Zone'. *Environmental Science and Technology.* 48(15): 65–80; Amankwaa, E. (2013). 'Livelihoods in Risk: Exploring the Health and Environmental Implications of E-Waste'. *Journal of Modern African Studies.* 51(4): 551–80; Wu, K. (2012). 'Electronic Waste Recycling and Neonatal Health Outcomes'. *Environment International.* 48: 1–8; UNEP (2011). *Where Are WEee in Africa? E-Waste Africa Programme* (UNEP, Nairobi). 8–9; Chi, X. (2011). 'Informal Electronic Waste Recycling in China'. *Waste Management.* 31:

The way the Basel Convention deals with this is by the assumption that if an old computer is to be reused, in most instances, it can be traded internationally as it is not hazardous waste. However, if it is to be recycled, it may be hazardous waste, and strong caution is urged in any potential international trade, by which prior informed consent must be obtained before any trade takes place.[78] To help further, the Parties have

---

731–42. UNEP (2011). *Study on the Possible Effects on Human Health and the Environment in Asia and the Pacific of the Trade of Products Containing Lead, Cadmium and Mercury*. (UNEP, Geneva). 12; Yu, J. (2010). 'Managing E-Waste in China'. *Resources, Conservation and Recycling*. 54: 991–9; Jiang, J. (2010). 'Electronic-Waste Recycling Plants'. *Journal of Environmental Sciences*. 22(5): 723–30; Sepulveda, A. (2010). 'Hazardous Substances Released from Electronic Equipment During Recycling'. *Environmental Impact Assessment Review*. 30: 28–41; Wath, S. (2010). 'Sustainable E-Waste Management'. *Science of the Total Environment*. 409: 19–32; Sinha, D. (2005). 'A Comparison of Electronic Waste Recycling'. *Environmental Impact Assessment Review*. 25: 492–504; Widmer, R. (2005). 'Global Perspectives on E-Waste'. *Environmental Impact Assessment Review*. 25: 436–58; Chen, D. (2009). 'Pollution Characterization of an E-Waste Dismantling Region'. *Environmental Pollution*. 157: 1051–57; Huo, Y. (2008). 'Toxic Heavy-Metal Waste Exposure and Abnormal Birth Outcomes'. *Toxicology Letters*. 180S: S32–S246; Wong, M. (2007). 'Uncontrolled Electronic-Waste Recycling'. *Environmental Pollution*. 149: 131–40; Leung, J. (2006). 'Environmental Contamination from Electronic Waste Recycling'. *Journal of Material Cycles Waste Management*. 8: 154.

[78] E-waste is included in Annex VIII of the Basel Convention, under notation A1180, whereby electronic waste will be assumed to be hazardous if it contains hazardous components also listed in Annex VIII such as, inter alia, lead (Y31); nickel-cadmium batteries (Y26); selenium (Y25); printed circuit boards (A1020/Y20); beryllium, beryllium compounds (Y27); antimony, antimony compounds (Y45); fluorescent tubes and backlight lamps from liquid crystal displays, which contain mercury (Y29); brominated flame retardants (linked to A3180, Y27 and Y45); waste oils/liquids (Y8 and Y9); and components containing asbestos (Y36). If any of the e-waste meets these criteria, and is therefore considered hazardous, it is essential that, assuming there is no national prohibition in that particular trade, for the trade to be lawful, the prior informed consent and additional considerations required by the Basel Convention be followed. E-waste is also included in Annex IX of the Basel Convention as non-hazardous wastes under notation B1110. This notation covers electrical and electronic assemblies, electronic assemblies consisting only of metals or alloys, waste electrical and electronic assemblies or scrap not containing components such as those listed above under A1180. For such waste that is not considered to be hazardous the Basel Convention does not foresee a specific procedure, although individual countries may have their own rules about this. For discussion on the reuse versus

begun to formulate guidelines for importing Parties on how to deal with this waste. However, at the time of writing, only draft guidelines exist.[79]

The final example where international law has attempted to set standards with regard to the recycling of metal is for ship-breaking. This is necessary because, when ship-breaking is done in poorly regulated environments, problems arise covering everything from the stripping of paint through to dealing with the heavy metals, asbestos and other toxic pollutants that remain within the vessel.[80] It was due to these problems, and the impacts they have upon workers and the environment alike, that the 2009 International Convention for the Safe and Environmentally Sound Recycling of Ships was concluded.[81] This Convention, assembled through the tripartite work of the International Labour Organization, the International Maritime Organization and the Basel Convention, applies to all non-military vessels above 500 tons that are to be recycled. This regime establishes controls relating to ship recycling in terms of both the vessels recycled and the facilities that this is done through. The goal is to:

> Prevent, reduce, minimize and, to the extent practicable, eliminate accidents, injuries and other adverse effects on human health and the environment

---

recycling approach on e-waste, see Li, J. (2013). 'Regional or Global WEEE Recycling: Where to Go?'. *Waste Management*. 18: 1–30; Basel Convention (2012). *Draft Technical Guidelines on Transboundary Movements of Electronic and Electrical Waste, In Particular, the Distinction Between Waste and Non-Waste*. (Basel Convention Secretariat, Basel). 5–6; UNEP (2011). *Where Are WEee in Africa? E-Waste Africa Programme*. (UNEP, Nairobi). 5–7, 20–22; UNEP (2011). *Study on the Possible Effects on Human Health of the Trade of Products Containing Lead, Cadmium and Mercury*. (UNEP, Nairobi). 12.

79 Basel Convention (2012). *Draft Technical Guidelines on Transboundary Movements of Electronic and Electrical Waste, In Particular, the Distinction Between Waste and Non-Waste*. (Basel Convention Secretariat, Basel).

80 Anon. (2012). 'Hard to Break Up'. *Economist*. 27 Oct. 26; Deshpande, P. (2012). 'Heavy Metal Exposure to Ship Recycling Yard Workers'. *Science of the Total Environment*. 438: 304–11; Chang, Y. (2010). 'Ship Recycling and Marine Pollution'. *Marine Pollution Bulletin*. 60: 1390–6; Paul, K. (2004). 'Exporting Responsibility: Shipbreaking in South Asia'. *Environmental Policy and the Law*. 34(2): 73; Clapp, J. (2001). *Toxic Exports*. (Cornell UP, NYC). 101–2.

81 For commentary, see Matz, L. (2010). 'Safe and Sound Scrapping of "Rusty Buckets"?' *RECIEL*. 19(1): 95–104; Harrison, J. (2009). 'The International Convention for the Safe and Environmentally Sound Recycling of Ships'. *The International Journal of Marine and Coastal Law*. 24: 727–36; Moen, E. (2008). 'Breaking Basel: The Elements of the Basel Convention and Its Application to Toxic Ships'. *Marine Policy*. 32: 1053–62.

> caused by ship recycling, and enhance ship safety, protection of human health and the environment throughout a ship's operating life.[82]

To ensure this goal, each ship must be surveyed and certified. Only certified ships going through authorised facilities can be recycled. If a ship is transferred, in violation of the agreed procedures (such as with inventory), it may be detained, dismissed or denied entry to the recycling facility. In addition, those responsible for the violation shall be prosecuted.[83] Each ship recycling facility must be authorised and closely regulated. Broadly, each facility must adopt management systems which do not pose health risks to the workers concerned or to the population in the vicinity of the recycling areas. The procedures must also prevent, reduce, minimise and, to the extent practicable, eliminate adverse effects on the environment. To achieve this, all ship recycling facilities must have detailed plans for both worker safety and environmental protection. These plans must cover the creation of safe working conditions, the provision of specified safety equipment, training of workers, emergency preparedness and response plans, record keeping and reporting. With regard to the dealing with hazardous wastes involved with ships, the process begins, for both new and those about to be recycled, with the requirement for a certificate on Inventory of Hazardous Materials, in which a number of hazards to be cognizant of are clearly identified. In addition, every authorised recycling facility must show it has the capacity to deal with (removal, storage and process) a series of listed hazards, such as the traditional heavy metals, persistent organic pollutants, asbestos, and ozone depleting substances. To achieve this, the guidelines for all of the recognised hazards must be taken into account.[84]

## 8. CONCLUSION

Recycling is the most common of environmental responses for many citizens and private enterprise. It is a well-recognised goal in domestic, regional and international law and policy, with roots going back hundreds of years. For most developed countries, the primary justification for

---

[82] Article 1.

[83] Articles 3, 4, 5, 9, 10 and 11. Also, regulation 8.

[84] See articles 6 and 8, and regulations 3, 5, 10, 15 to 23, as well as Appendices 2 and 5. Supplemental, and very detailed, guidelines on the facility, to ensure safe and environmentally friendly recycling, had to be developed over the following years. See MEPC (2011). 'Recycling of Ships: Guidelines for Safe and Environmentally Sound Ship Recycling'. MEPC 62/3. 8 Apr.

recycling in the 21st century is environmental, but in all instances, the economic context provides a very solid foundation. This context of recycling is evident in all waste streams, with some waste streams showing great success at recycling, whilst in others, such as paper, glass, rubber and biowaste, there is limited uptake of recycling, and the markets remain sluggish, often only existing because of governmental support.

The situation is different with the markets involving recycling metals (from ships to electronic waste) and some plastics. In these instances, despite clear environmental benefits, the profit margins alone are driving the industries. The problem is that these markets for recycling, which are largely international, can be driven so aggressively that they may end up causing significant environmental and/or social damage to the people engaged in the recycling and their surrounding environment. This is especially so when the recycling industries operate in informal contexts and the countries are willing to tolerate degrees of damage in the pursuit of the economic benefits of recycling, that more developed countries are not. Due to such risks, the international community has developed a series of rules and guidelines, which are designed to mitigate the worst impacts of these recycling trades. However, it is premature to say if the intentions of international law in this area, to make recycling safer and more environmentally sound, will win out against the profit motive in all recycling sectors and the desire to keep recycling costs as low as possible.

# 5. Disposal into the air

## 1. INTRODUCTION

The burning of waste in open spaces is a practice which goes back thousands of years. Today, it remains a common form of waste disposal in the developing world. Conversely, it is uncommon in developed countries, and especially in urban areas, to light fires to burn waste in an unregulated manner. This practice has largely come to an end because of the clear environmental and social risks of poor incineration and the associated air pollution that is imposed upon surrounding communities when waste is burnt in an ineffective way. These hazards have increased as the waste streams have changed, and the waste that is burnt is more likely to include material containing toxic contaminants.

This is not to suggest that the burning of waste does not happen in the developed world. This is incorrect. What has largely ended is uncontrolled burning. The practice of controlled burning, via either industrial incinerators, cement kilns or medical incinerators, is an ongoing practice.

Where debate has arisen has been with the incineration of municipal waste for the twin purposes of waste disposal and energy recovery. This is an issue which evolved out of the 1990s when it became argued that burning waste to dispose of it, and make energy from it at the same time, may be the best option. The opposing view has been that controlled burning, like open burning, may also release dangerous pollutants. The surrounding evidence is the source of debate. This is especially so when the restrictions on what can be burnt have dovetailed with strong standards for the efficiency of burning and the adequacy of pollutant capture.

International law, in both specific and general terms, supports the incineration of waste streams on land, but not at sea, with the latter being prohibited in 1996 after 20 years of ever increasing standards. On land, through multiple national settings, the imposition of high standards appears to have mitigated many of the impacts associated with incineration in earlier decades. However, this appearance only exists for the countries which possess the high standards and the state of the art technology to achieve them. In countries without these standards, it is possible that the inefficient burning of waste is a primary source of air pollution.

## 2. OPEN BURNING

For fire to burn effectively, a sufficient mixing of fuel and oxygen must occur. This should generate maximum amounts of combustion, via an efficient burn of the fuel at hand. In theory, this should create more heat and cause greater destruction of what is being burnt. The greater the temperature, the more will be destroyed, including many types of different pollutants. If something is not burnt efficiently, it will generate less heat and more pollutants. This is a particular problem with what is known as 'open burning'.

Open burning is where unenclosed combustion of materials in an ambient environment is undertaken. The advantage of open burning is that it reduces the amount of waste in any given location, and in turn, many of the nuisances, from vermin to odour, associated with it. The first evidence of collecting and burning waste in the open comes from Crete around 2000 BCE, and by the year 1200 BCE, the Old Testament recorded the people of Jerusalem burning some of their waste in fires emanating from natural gas vents in the nearby Valley of Gehenna, which through a process of association, became a synonym for 'hell'. Some 3,200 years later, open burning of waste remains a common practice in many developing countries. It also still exists in pockets of the developing world, such as some rural environments or in some small scale domestic fires in the backyards of citizens.[1]

Open burning, due to its less than ideal combustion conditions, rarely achieves the goals of safe combustion. Inefficient packing, low heat, limited airflow and smouldering – all characteristics of open burning – are ideal for the creation of more levels of pollutants than when waste is burnt effectively. The damage that is done from such burning depends upon what is burnt, how much is burnt, how it is burnt and what sensitive ecosystems or creatures are nearby. This problem applies to the incineration of both organic and inorganic waste. The burning of organic waste has always been a problem if it is not entirely dry. Conventional air pollution in the forms of particulate matter (of the larger size – PM10 –

---

[1] UNEP (2012). *Global Environment Outlook 5*. (Progress Publishers, Malta). 173; Rathje, W. (1992). *Rubbish: The Archaeology of Garbage*. (HarperCollins, NYC). 19.

not the more micro size associated with industrial processes), via mixes of carbon, sulphur and nitrogen, are all common. However, as a source of air pollution and climate change, these have traditionally been under-represented in official statistics. However, this is beginning to change as hazardous pollutants from open burning are increasingly becoming recognised as a dominant source of emissions worldwide. Studies show that the source of open burning of waste and the creation of toxic pollutants may (depending on what is burnt) be up to 2,000 times higher than those from stacks of modern municipal waste combustors. Accordingly, for some pollutants such as dioxins and furans (PCDD/Fs) open burning can account for the dominant share of national emissions.[2] Similarly, in China, the emissions of particulate matter from open waste burning is the equivalent of 22 percent of their total reported anthropogenic particulate matter emissions.[3]

The open burning of inorganic waste and new types of pollution are a much more modern problem that has grown as new waste streams have evolved. Here, burning, inter alia, tyres, chlorinated plastics, electronic equipment, consumer products caked in flame retardants, and/or old pesticides will each release an entirely different stream of pollutants. Depending on what is burnt, carbon monoxide, methane and volatile organic compounds (such as benzene and polycyclic aromatic compounds) may emerge. Depending on the source, varying amounts of metals such as lead or mercury may also be emitted. Potentially

---

[2] Bastian, L. (2013). 'Behavior of PCDD/Fs During Open Burning of Municipal Solid Waste in Open Dumping Sites'. *Journal of Material Cycles and Waste Management*. 15: 229–41; Aurell, J. (2012). 'Emissions from Open Burning'. *Environmental Science and Technology*. 46: 11004–12; Solorzano, G. (2012). 'Open Burning of Household Waste: Effect of Experimental Condition on Combustion Quality and Emission of PCDD, PCDF and PCB'. *Chemosphere*. 87: 1003–1008; Forbid, G. (2011). 'Open Waste Burning in Cameroonian Cities: An Environmental Impact Analysis'. *Environmentalist*. 31: 254–62; Zhang, T. (2011). 'Emissions of Unintentional Persistent Organic Pollutants from Open Burning of Municipal Solid Waste from Developing Countries'. *Chemosphere*. 84: 994–1001; Gullett, B. (2010). 'PCDD/F, PBDD/F, and PBDE Emissions from Open Burning of a Residential Waste Dump'. *Environmental Science and Technology*. 44: 394–9; Sidhu, S. (2005). 'Endocrine Disrupting Chemical Emissions from Combustion Sources: Diesel Particulate Emissions and Domestic Waste Open Burn Emissions'. *Atmospheric Environment*. 39. 801–11.

[3] Wiedinmyer, C. (2014). 'Global Emissions of Trace Gases, Particulate Matter and Hazardous Air Pollutants from Open Burning of Domestic Waste'. *Environmental Science and Technology*. 48(16): 9523–30; Gillespie, A. (2006). *Climate Change, Ozone Depletion and Air Pollution*. (Brill, The Netherlands). Chapter 2.

endocrine disrupting chemicals can be produced, as can persistent organic pollutants. Of the latter, polychlorinated dibenzo-p-dioxins and polychlorinated dibenzofurans (PCDDs/Fs) or polychlorinated biphenyls (PCBs) are the most common (depending on the waste stream) and can also be emitted in disproportionately large amounts.[4]

## 3. CONTROLLED BURNING

On the whole, in most developed countries, open burning is a practice of the past. Controlled burning, however, is not. Open burning became controlled burning in 1874 when the first modern garbage incinerator, known as a 'destructor', went into operation in Nottingham, England. Here, for the first time, attempts were made to make the fire hotter by feeding it continuously with a steady stream of combustible material. Within a decade the technology had spread to the United States where it proliferated quickly, with more than 180 furnaces being erected between 1885 and 1908. After the Second World War, over 700 incinerators were in operation in the United States and Europe. By 1960, almost a third of municipal waste in the United States was being incinerated.[5] Around the same time period, it also became apparent that some kinds of hazardous wastes such as persistent organic pollutants, PCBs, medical waste,

---

4 Luo, Q. (2013). 'Polybrominated Diphenyl Ethers in Combusted Residues and Soils from an Open Burning Site of Electronic Wastes'. *Environmental Health Sciences*. 69(8): 2633–45; Bastian, L. (2013). 'Behavior of PCDD/Fs During Open Burning of Municipal Solid Waste in Open Dump Sites'. *Journal of Material Cycles and Waste Management*. 15(2): 229–41; Lundin, L. (2013). 'The Effect of Developing Nations' Municipal Waste Composition on PCDD/PCDF Emissions from Open Burning'. *Atmospheric Environment*. 79: 433–41; Zhang, T. (2011). 'Emissions of Unintentional Persistent Organic Pollutants from Open Burning of Municipal Solid Waste from Developing Countries'. *Chemosphere*. 84: 994–1001; Forbid, G. (2011). 'Open Waste Burning in Cameroonian Cities: An Environmental Impact Analysis'. *Environmentalist*. 31: 254–62; Estrellan, C. (2010). 'Toxic Emissions from Open Burning'. *Chemosphere*. 80: 193–207; Lemieux, P. (2004). 'Emissions of Organic Air Toxics from Open Burning: A Comprehensive Review'. *Progress in Energy and Combustion Science*. 30: 1–32; Wevers, R. (2004). 'Effect of Backyard Burning on Dioxin Deposition and Air Concentrations'. *Chemosphere*. 54: 1351–6.

5 Walsh, D. (2002). 'The Evolution of Refuse Incineration'. *Environmental Science and Technology*. 36(15): 316–24; Colten, C. (1996). *The Road to Love Canal: Managing Industrial Waste Before the EPA*. (Austin, Texas UP). 54–7, 63; Garrett, E. (1890). *Law of Nuisances*. (Clowes, London). 270–75.

sewage sludge and mineral oils could be safely destroyed by carefully managed, high-standard incineration.[6]

Finally, it became apparent that incineration[7] could not only destroy some waste streams, but also, in the process of doing so, it could generate energy which could be harvested or 'recovered'. This became apparent in the 1960s when a number of industrial plants that required large amounts of energy to achieve ongoing supplies of very high temperatures, such as steel plants and cement kilns, began to look at the incineration of waste as an alternative or supplementary fuel source for which some waste streams proved ideal, such as used tyres, which has become the fate for half of this particular rubber waste stream in the United States.[8]

Later research showed that the heat from incinerating waste could also be captured for energy purposes, via producing steam which turns turbines to generate electricity. The basic equation suggested that 1 ton of municipal rubbish (especially if it contains high energy waste like

---

[6] Lin, H. (2012). 'Simulation of Co-Incineration of Sewage Sludge with Municipal Solid Waste'. *Waste Management.* 32(3): 561–7; Chen, T. (2012). 'Fixation and Partitioning of Heavy Metals in Slag after Incineration of Sewage Sludge'. *Waste Management.* 32(5): 957–64; Roy, M. (2011). 'Review of Bio-Solids Management Options and Co-Incineration'. *Waste Management.* 31(11): 2228–35; UN-Habitat (2008). *Global Atlas of Excreta, Wastewater Sludge, and Biosolids Management.* (Habitat, Nairobi). 27–9; Choi, C. (2004). 'Power Sludge'. *Scientific American.* May. 20.

[7] Although it is possible that waste may be turned into energy from other processes in time to come, for the foreseeable future, it is likely that incineration will be the primary conduit for this type of energy recovery. The contending non-incineration technologies for energy recovery in the future include hydrogenation, pyrolysis, gasification and/or bioconversion. See Fador, Z. (2012). 'Waste as Alternative Fuel'. *Process Safety and Environmental Protection.* 9: 263–84; Thanh, N. (2012). 'Household Solid Waste Treatment Practices Using Life Cycle Inventory Assessment Mode'. *Environmental Monitoring and Assessment.* 184: 3515–27; Demirbas, F. (2011). 'Biowastes-to-Biofuels'. *Energy Conversion and Management.* 52: 1815–28.

[8] Karstensen, K. (2010). 'Test Burn with PCB-Oil in a Local Cement Kiln'. *Chemosphere.* 78(6): 717–23; Smith, C. (2009). 'Lesser Known Energy Sources: A Study of Biogas and Tire-Based Fuel'. *Cogeneration & Distributed Generation Journal.* 23(2): 77–91; Caveny, B. (2008). 'Tires Fuel Cement Manufacturing'. *Oil and Gas Journal.* 96(35). 64–70; Anon. (2007). 'Nova Scotia Chooses Cement Kiln for Tires'. *Solid Waste & Recycling.* 12(2): 30; Gue, R. (2006). 'The Cement Kiln Option'. *Solid Waste & Recycling.* 11(3): 8–12; Nicholson, J. (2005). 'The Tire Recycling Market'. *Solid Waste & Recycling.* 10(5): 42–3; Mokrzycki, E. (2003). 'Use of Alternative Fuels in the Cement Industry'. *Applied Energy.* 74(1): 101–11.

oil-based plastic or rubber, paper or dried sludge) produces as much heat energy as 250 kilograms of coal. The additional benefit of this was not only that it was burning a very cheap fuel source, it was also reducing demand for the alternative fuels, which in itself could create environmental benefits and provide energy security.[9] These views were first recognised by the United States Congress with their 1976 Resource Recovery Act,[10] which recognised that:

> [S]olid waste represents a potential source of solid fuel, oil, or gas that can be converted into energy; ... the need exists to develop alternative energy sources for public and private consumption in order to reduce our dependence on such sources as petroleum products, natural gas, nuclear and hydroelectric generation; and technology exists to produce usable energy from solid waste.[11]

Nearly 40 years later, the Europeans came to a similar conclusion and came to endorse waste-to-energy incinerators as operations of value, involved in the 'recovery' of materials, rather than simply being waste disposal operations. This change of classification brought a change of financial support.[12] The difference with the Europeans was that, by the end of the first decade of the 21st century, it was also being argued that

---

[9] Passamonti, F. (2012). 'Recycling of Waste Plastics into Fuels'. *Applied Catalysis B: Environmental*. 125: 499–506; Briassoulis, D. (2012). 'The Quality Characteristics of Agricultural Plastic Wastes Regarding their Recycling and Energy Recovery Potential'. *Waste Management*. 32: 1075–90; Chen, W. (2011). 'The Potential Environmental Gains from Recycling Waste Plastics'. *Waste Management*. 31: 168–79; Zhang, G. (2007). 'Prospect and Current Status of Recycling Waste Plastics and Technology for Converting them into Oil in China'. *Resources, Conservation and Recycling*. 50: 231–9; Porteous, A. (2005). 'Why Energy from Waste Incineration is an Essential Component of Environmentally Responsible Waste Management'. *Waste Management*. 25: 451–9; Porteous, A. (2001). 'Energy from Waste Incineration: A State of the Art Emissions Review'. *Applied Energy*. 70: 157–67; Gelfand, L. (2001). 'Waste to Energy Incineration'. *Energy Engineering*. 98(1): 23–46.

[10] Resource Recovery Act. 84 Stat. 1227 1970–1971. Also, Public Law 91-512-26 Oct. 1970. Section 205. Also, Rathje, W. (1992). *Rubbish: The Archeology of Garbage*. (Harper, NYC). 174–8; Rogers, H. (2005). *Gone Tomorrow: The Hidden Life of Garbage*. (Norton, NYC). 68–70, 80–81, 159; Porter, R. (2002). *The Economics of Waste*. (Resources for the Future, Washington). 71.

[11] Resource Recovery Act. 90 Stat. 2795 1976. Also, Public Law 94-580-21 Oct. 1976. Section 1002(d).

[12] Pavlas, M. (2011). 'Waste Incineration with Production of Clean and Reliable Energy'. *Clean Technologies and Environmental Policies*. 13(4): 954–60; Grosso, M. (2010). 'Efficiency of Energy Recovery from Waste Incineration

the incineration of many types of waste was actually the most environmentally friendly disposal option. The theoretical thinking behind this contention was the 'life-cycle approach'. The life-cycle approach involves every part of the cycle of a product or process. The goal is to ensure that the least environmental impact, in both direct and indirect terms, is achieved. The attempt to reconcile these has been through life-cycle analysis of each of the disposal options (in terms of both direct and indirect costs) of the same waste.[13] This thinking was very persuasive, for as the international community concluded at the 2012 'Rio+20' conference:

> We recognize the importance of adopting a life cycle approach and of further development and implementation of policies for resource efficiency and environmentally sound waste management. We therefore commit to further reduce, reuse and recycle waste … with a view to managing the majority of global waste in an environmentally sound manner and, where possible, as a resource.[14]

---

in Light of the New Waste Framework Directive'. *Waste Management.* 30(7): 1238–48.

[13] Di Maria, F. (2014). 'A Holistic Life Cycle Analysis of Waste Management Scenarios'. *Waste Management.* 34(11): 2382–90; Othman, T. (2013). 'Review on Life Cycle Assessment of Integrated Solid Waste Management'. *Journal of Cleaner Production.* 41: 251–62; Zaman, A. (2013). 'The Zero Waste Index'. *Journal of Cleaner Production.* 50: 123–32; Koroneus, T. (2012). 'Integrated Solid Waste Management and Energy Production: A Life Cycle Assessment Approach'. *Journal of Cleaner Production.* 27: 141–50; Ulgiati, S. (2011). 'Material, Energy and Environmental Performance of Technological and Social Systems under a Life Cycle Assessment Perspective'. *Ecological Modelling.* 222: 176–89; Turconi, R. (2011). 'Life Cycle Assessment of Waste Incineration in Denmark and Italy Using Two LCA Models'. *Waste Management & Research.* 29(10): 78–90; Massarutto, A. (2011). 'Material and Energy Recovery in Integrated Waste Management Systems: A Life-Cycle Costing Approach'. *Waste Management.* 31: 2102–11; Gentil, E. (2010). 'Models for Waste Life Cycle Assessment: Review of Technical Assumptions'. *Waste Management.* 30: 2636–48; Wittmaier, M. (2009). 'Possibilities and Limitations of Life Cycle Assessment in the Development of Waste Utilization Systems'. *Waste Management.* 29: 1732–8; Winkler, J. (2007). 'Comparative Evaluation of Life Cycle Assessment Models for Solid Waste Management'. *Waste Management.* 27: 1021–31; Clift, I. (2000). 'The Application of Life Cycle Assessment to Solid Waste Management'. *Transactions of the Institute of Chemical Engineers.* 78(B): 280–314.

[14] RIO+20. A/CONF.216/XX. Paragraph 218.

The cumulative result of all of this support for the controlled incineration of waste is that, globally, the operations expanded quickly during the last decades of the 20th century. On the high seas, before it was prohibited at the end of the century, incineration went from zero to 200,000 tons per year, between 1970 and 1990.[15] On land, the incineration of waste has increased from 160 million tons per year in 1990 to an estimated 240 million tons per year by 2016. This growth is particularly obvious in countries which are short on space, and/or the economic costs or legal possibilities of other forms of disposal are not attractive. For countries such as Singapore and Japan, more than 50 percent of their waste is incinerated. In the United States, 87 incinerators burn 11.7 percent of all the municipal waste. This burning is supplemented by some 26 cement kilns which also burn waste. In Europe, there are over 400 incinerators, with around another 50 in various phases of construction. In China, high-tech incineration has grown from zero plants in 1980 to 70 in 2010. These currently combust some 33,000 tons of municipal waste per day. These numbers in China are projected to increase rapidly in the coming decades.[16]

---

15 IMO (1991). *The London Dumping Convention: The First Decade.* (IMO, London). 18; Anon. (1988). 'Incineration Ships Poison North Sea's Sediments'. *New Scientist.* 29 Sept. 33; Caufield, C. (1984). 'More Toxic Waste to Be Burned at Sea'. *New Scientist.* 12 Apr. 10; Anon. (1984). 'Sea Burn'. *New Scientist.* 3 May. 5; Anon. (1984). 'Toxic Burn Ban'. *New Scientist.* 7 June. 5; Office of Technology Assessment (1981). *Ocean Incineration: Its Role in Managing Hazardous Waste.* (OTA, Washington).

16 Buekens, D. (2011). 'Waste Incineration, PVC, and Dioxins'. *Journal of Material Cycles Waste Management.* 13: 190–97; EPA (2011). *Municipal Solid Waste Generation, Recycling, and Disposal in the United States: Facts and Figures for 2010.* (EPA, Washington, EPA-530-F-11-005). 3–4; Yan, M. (2011). 'Persistent Organic Pollutant Emissions from Medical Waste Incinerators in China'. *Journal of Material Cycles Waste Management.* 13: 213–18; Saner, D. (2011). 'Scenarios for the Implementation of EU Waste Legislation at National Level and their Consequences for Emissions from Municipal Waste Incineration'. *Resources, Conservation and Recycling.* 57: 67–77; Zhang, D. (2010). 'Municipal Solid Waste Management in China'. *Journal of Environmental Management.* 91(8): 1623–33; Anon. (2010). 'The Worldwide Market for Waste Incineration Plants'. *Ecology, Environment & Conservation Business.* 23 Jan. 362; Grosso, M. (2010). 'Efficiency of Energy Recovery from Waste Incineration'. *Waste Management.* 30: 1238–43; Pellow, D. (2007). *Resisting Global Toxics.* (MIT Press, Cambridge). 99, 105; Porter, R. (2002). *The Economics of Waste.* (Resources for the Future, Washington). 71; Charles, D. (1997). 'Recycling Madness Revives Obsolete Plants'. *New Scientist.* 8 Mar. 6; Pearce, F. (1997). 'Burn Me'. *New*

## 4. THE RISKS OF CONTROLLED BURNING

Despite the growth in such waste disposal technologies, the practices have been (and are) highly contested. The basic problem for controlled burning is exactly the same as it is for open burning. Inefficient incineration may cause pollutants which can cause environmental and/or social harm. This problem became apparent in the 1970s when studies began to show that the inefficient incineration of plastic, sewage sludge and/or hazardous wastes such as obsolete pesticides could result in the production of the same harmful substances that were also often associated with open burning. In particular heavy metals, mercury and/or organic compounds like halogenated dioxins, furans and PCDDs/Fs were being located in the emissions of many incinerators (and similar industrial waste combustion facilities, like cement kilns). By the year 2000, municipal incinerators were attributed for creating, inter alia, 69 percent of worldwide emissions of dioxin. The problem was that many of the hazards in the waste were not being effectively destroyed in incineration due to inappropriate temperature, turbulence and/or burning time. Because of such failures, the pollutants either passed through the combustion process or morphed into new types of pollutants.[17]

---

*Scientist*. 22 Nov. 20–24; Kim, I. (1994). 'Incinerators and Cement Kilns Face Off'. *Chemical Engineering*. 101: 41–50.

[17] Sipkova, A. (2014). 'Mercury Distribution and Mobility in Contaminated Soils from Vicinity of Waste Incineration Plant'. *Plant, Soil and Environment*. 60(2): 87–92; Takaoka, M. (2012). 'Mercury Emission from Sewage Sludge Incineration in Japan'. *The Journal of Material Cycles and Waste Management*. 14(2): 113–19; Chen, T. (2012). 'Fixation and Partitioning of Heavy Metals in Slag after Incineration of Sewage Sludge'. *Waste Management*. 32(5): 957; Biganzoli, L. (2012). 'Chemical and Sewage Sludge Co-Incineration'. *Waste Management Research*. 30(10): 1081–8; Murakami, T. (2009). 'Combustion Characteristics of Sewage Sludge in an Incineration Plant for Energy Recovery'. *Fuel Processing Technology*. 90(6): 778–83; Helge, K. (2008). 'Formation, Release and Control of Dioxins in Cement Kilns'. *Chemosphere*. 70(4): 543–60; Bui, V. (2007). 'New Technique for Improvement on Reactivity of Cement Industry Wastes'. *Advances in Applied Ceramics*. 106(5): 226–30; Anon. (2007). 'PM Standard for New Cement Kilns that Burn Hazardous Waste Revised'. *Air Pollution Consultant*. 17(1): 51–7; Pellow, D. (2007). *Resisting Global Toxics*. (MIT Press, Cambridge). 103; Stasta, P. (2006). 'Thermal Processing of Sewage Sludge'. *Applied Thermal Engineering*. 26(13): 1420–26; Anon. (2004). 'Incinerators Linked to Twin Births'. *New Scientist*. 12 June. 17; Lee, M. (2003). 'Economics: Waste Recycling'. *The Ecologist*. 33(4): 50–51; Werter, J. (1999). 'Sewage Sludge Combustion'. *Progress in Energy and Combustion Science*. 25: 55–116; Anon. (1998). 'PCB Pollution'. *New Scientist*. 29 Jan. 27; MacKenzie,

Such emissions from inefficient industrial incinerators and/or related technologies, like cement kilns, have been linked to detrimental impacts on human health for communities living nearby. In this area, a number of studies have found evidence of negative pregnancy outcomes (including increased risks for central nervous system defects, congenital heart defects, oral clefts, renal dysplasia, limb malformations, chromosomal anomalies, preterm births, low birth weight and infant deaths), childhood cancers, asthma hospitalisations and chronic respiratory symptoms, heart disease and diabetes.[18]

The difficulties with such findings are twofold. First, problems of cause and effect plague this area. That is, whilst chemicals like dioxins are a known human carcinogen, most of the evidence linking these to cancer in humans is based on studies of populations that have been exposed to high concentrations of dioxins either occupationally or through industrial accidents. Evidence linking dioxins and furans to cancer in humans in which there is low-level exposure, such as residence near an incinerator, even if prolonged, is less secure. Specifically, a number of studies suggest that there is no clear evidence linking high-tech incineration plants to the creation of pollutants which harm people or the environment. This contra view argues that many studies in this area involve incomplete cohorts, bias, lack of data on individual exposure, and most importantly, too many other explanations for where

---

D. (1997). 'Burning Plastic'. *New Scientist.* 3 May. 4; Hecht, J. (1991). 'Incinerators Blamed for Toxic Fallout'. *New Scientist.* 7 June. 14; Anon. (1990). 'Dioxin Cuts'. *New Scientist.* 5 May. 9; Anon. (1989). 'Confusion over PCBs'. *New Scientist.* 19 Aug. 7; Milne, R. (1988). 'Hospital Incinerators Stoke up Pollution Row'. *New Scientist.* 30 June. 38; Pearce, F. (1985). 'Incinerators and Birth Defects'. *New Scientist.* 21 Feb. 4; Anon. (1984). 'Where Dioxin may Deform in Britain'. *New Scientist.* 26 July. 8; Anon. (1985). 'Dioxin Fears'. *New Scientist.* 3 Jan. 5; Anon. (1985). 'PCBs Crisis'. *New Scientist.* 12 Sept. 20; Anon. (1985). 'Closure of Incinerator'. *New Scientist.* 12 Sept. 20; Anon. (1984). 'PCB Incinerator Closes'. *New Scientist.* 5 Sept. 5.

[18] Brender, N. (2011). 'Residential Proximity to Environmental Hazards and Adverse Health Outcomes'. *American Journal of Public Health.* 101 (Supp 1): 12–20; Gautam, V. (2010). 'Biomedical Waste Management: Incineration vs. Environmental Safety'. *Indian Journal of Medical Microbiology.* 28(3): 191–210; Keane, M. (2007). 'Catalytic Conversion of Waste Plastics'. *Journal of Chemical Technology and Biotechnology.* 82: 787–95; Thompson, J. (2005). 'The Health Effects of Waste Incinerators'. *Journal of Nutritional & Environmental Medicine.* 15(2/3): 115–56; Cordier, S. (2004). 'Risk of Congenital Anomalies in the Vicinity of Municipal Solid Waste Incinerators'. *Occupational Environmental Medicine.* 61: 8–15; Klein, K. (1996). 'Long Lived Pollutants Threaten the Great Lakes'. *New Scientist.* 13 July. 12.

the environmental pollutants may have come from. Second, and perhaps most fundamentally of all, each study is highly dependent upon what is burnt and how well it was incinerated. That is, when suitable pollution prevention technologies and enhanced efficiencies in technology are implemented, levels of pollutants are greatly reduced. Thus, the likelihood of pollutants coming from a controlled burn incinerator plant is directly related to the standards of the plant in terms of, inter alia, pollution traps, what is burnt and how, and how close the surrounding population is to the location.[19]

A less immediate problem, but one of enduring concern, is the ash residue of the incineration process. This is usually about 10 percent of the weight of what was burnt. There are two types of ash with modern incineration processes. These are fly ash (which follows waste gases through the exhaust and scrubbing systems) and bottom ash (which is heavier and falls through grates). Both types of ash have the potential – depending on what is incinerated and how efficiently – to be very toxic, possessing, inter alia, heavy metals and dioxin. The dust from cement kilns can also contain hazardous residues.[20] Whilst some reuse of the ash or dust is possible in some instances, such as making it inert and using it for base industrial products like roading or cement, more often than not it has to be disposed of in specially designed landfills.[21] It was the desire to avoid the costs of putting such ash waste in a specially designed landfill

---

[19] Nzihou, A. (2012). 'Dioxin Emissions from Municipal Solid Waste Incinerators in France'. *Waste Management.* (32): 2273–7; Ranzi, A. (2011). 'Mortality and Morbidity among People Living Close to Incinerators'. *Environmental Health.* 10: 22; Vincenti, M. (2008). 'Adverse Pregnancy Outcomes in a Population Exposed to the Emissions of a Municipal Waste Incinerator'. *Science of the Total Environment.* 407: 116–21; Dummer, T. (2003). 'Adverse Pregnancy Outcomes around Incinerators and Crematoriums in Cumbria, North West England, 1956–93'. *Journal of Epidemiological Community Health.* 57: 456–61.

[20] Tzanakos, K. (2014). 'Solidification of Ash from Medical Waste Incineration'. *Waste Management.* 34(10): 1823–30; Tan, Z. (2012). 'Leaching Characteristics of Fly Ash from Medical Waste'. *Waste Management and Research.* 30(3): 285–92; Wang, W. (2010). 'Release of Elements from Municipal Solid Waste Incineration Fly Ash'. *Frontiers of Environmental Science Engineering.* 4(4): 482–9; Dabo, D. (2009). 'Ten-Year Chemical Evolution of Leachate and Municipal Solid Waste Incineration Bottom Ash'. *Journal of Hazardous Materials.* 172: 904–13; Sabbas, T. (2003). 'Management of Municipal Solid Waste Incineration Residues'. *Waste Management.* 23(1): 61–88; El-Awady, M. (1997). 'Removal of Heavy Metals by Cement Kiln Dust'. *Bulletin of Environmental Contamination and Toxicology.* 59(4): 603–10.

[21] Ahmari, S. (2013). 'Utilization of Cement Kiln Dust to Enhance Mine Tailings-Based Geopolymer Bricks'. *Construction and Building Materials.* 40:

by sending it overseas to countries with lesser standards (with the infamous waste ship *Khian Sea* in 1986) that created much of the emphasis for the formation of what became the Basel Convention.

A final problem with the advocacy for controlled incineration is, whilst few can disagree with the ideal of life-cycle approach, a final answer of what is the best waste disposal option for various types of waste has turned out to be a very complicated debate. This has been especially so around the issue of 'energy recovery'. Aside the areas where there is a clear answer (such as burning glass for energy which everyone agrees is a failure, as it just melts), with other waste streams, such as plastic, rubber and paper, the answer is not so clear-cut, as each of these has strong potential as an energy source – as well as recycling value. As such, it is not at all clear whether it is best to recycle these waste streams, or incinerate them and reap their energy.[22]

The problem is that life-cycle analysis for waste is not a harmonised science. The essence of this complication is the amount of factors and methodologies for both economic and ecological contexts that are put into the equation. When these methodologies are then extrapolated from

1002–11; Mohamedzein, Y. (2012). 'The Use of Municipal Solid Waste Incinerator Ash to Stabilize Dune Sands'. *Geotechnology, Geological Engineering*. 30: 1335–44; Zhongxin, T. (2011). 'Leaching Characteristics of Fly Ash from Chinese Medical Waste Incineration'. *Waste Management & Research*. 30(3): 285–94; World Bank (2012). *What a Waste: A Global Review of Solid Waste Management.* (World Bank, Urban Development Series, No. 15, Washington). 5; UN Human Settlements Programme (2011). *Solid Waste Management in the World's Cities*. (UNSP, NYC). xxviii; Ahmed, A. (2009). 'Use of Factory-Waste Shingles and Cement Kiln Dust to Enhance the Performance of Soil Used in Road Works'. *Advances in Civil Engineering*. 9–16; Lachemi, M. (2008). 'Controlled Low Strength Materials Incorporating Cement Kiln Dust'. *Cement and Concrete Composites*. 30(5): 381–92; Lin, C. (2006). 'Recovery of Municipal Waste Incineration Bottom Ash and Water Treatment Sludge to Water Permeable Pavement Materials'. *Waste Management*. 26: 970–78; Sabba, T. (2003). 'Management of Municipal Solid Waste Incineration Residues'. *Waste Management*. 23: 61–88; Sloot, H. (2001). 'Characteristics, Treatment and Utilization of Residues from Municipal Waste Incineration'. *Waste Management*. 21: 753–65; Al-Harthy, A. (2003). 'Effect of Cement Kiln Dust on Mortar and Concrete Mixtures'. *Construction and Building Materials*. 17(5): 353–60; Johannessen, K. (1996). 'The Regulation of Municipal Waste Incineration Ash: A Legal Review and Update'. *Journal of Hazardous Materials*. 47: 383–93.

[22] Anon. (2012). 'Less than Zero Waste'. *BioCycle*. 53(2): 10; Lombardi, E. (2011). 'Zero Landfill is Not Zero Waste'. *BioCycle*. 52(7): 44–5; Fiksel, K. (2011). 'Comparative Life Cycle Assessment of Beneficial Applications for Scrap Tires'. *Clean Technologies and Environmental Policy*. 13(1): 19–35.

and used for comparative analysis of different waste disposal options, fundamentally opposing conclusions can be arrived at. The foremost considerations that can change the answers in this area are the exact types of waste at issue, the pollutants that are being focused upon (and/or excluded), geographic and local particulars, the wages of workers, transportation costs, timescale, the specific technologies and their efficiencies, and associated standards, with particular regard to pollution control and labour standards. Differences between direct and indirect economic and environmental costs can complicate this matter even further. When overlaid with differentiated local, national, regional and/or international compliance considerations, the outcomes may twist yet again.[23] For example, with waste paper, the answer is heavily weighted by what types of post-consumer paper are being burnt, the incineration technology utilised, what energy is replaced by incinerated waste paper, what material is replaced by recycled fibres, how the pulpwood savings are used, and what environmental burdens are associated with a change in energy demand. When global warming considerations are added into the mix, the answer may change again. That is, when global warming emissions of incineration are added and the opportunity costs involving virgin wood sources, recycling usually (but not always) becomes the

---

[23] Galvez, J. (2014). 'An Analysis of the Use of Life Cycle Assessment for Waste Co-Incineration'. *Resources, Conservation and Recycling*. 86: 118–35; Othman, T. (2013). 'Review on Life Cycle Assessment of Integrated Solid Waste Management'. *Journal of Cleaner Production*. 41: 251–62; Zaman, A. (2013). 'The Zero Waste Index'. *Journal of Cleaner Production*. 50: 123–32; Koroneus, T. (2012). 'Integrated Solid Waste Management and Energy Production: A Life Cycle Assessment Approach'. *Journal of Cleaner Production*. 27: 141–50; Ulgiati, S. (2011). 'Material, Energy and Environmental Performance of Technological and Social Systems under a Life Cycle Assessment Perspective'. *Ecological Modelling*. 222: 176–89; Turconi, R. (2011). 'Life Cycle Assessment of Waste Incineration in Denmark and Italy Using Two LCA Models'. *Waste Management & Research*. 29(10): 78–90; Massarutto, A. (2011). 'Material and Energy Recovery in Integrated Waste Management Systems: A Life-Cycle Costing Approach'. *Waste Management*. 31: 2102–11; Gentil, E. (2010). 'Models for Waste Life Cycle Assessment: Review of Technical Assumptions'. *Waste Management*. 30: 2636–48; Wittmaier, M. (2009). 'Possibilities and Limitations of Life Cycle Assessment in the Development of Waste Utilization Systems'. *Waste Management*. 29: 1732–8; Winkler, J. (2007). 'Comparative Evaluation of Life Cycle Assessment Models for Solid Waste Management'. *Waste Management*. 27: 1021–31; Clift, I. (2000). 'The Application of Life Cycle Assessment to Solid Waste Management'. *Transactions of the Institute of Chemical Engineers*. 78(B): 280–314.

recommended option as having the better life-cycle analysis in terms of the lesser overall environmental impact.[24]

## 5. MANAGING THE RISK

Although most countries have facilities to incinerate medical waste, and active cement kilns, a number of countries, such as Australia and New Zealand, have expressly prohibited the incineration of municipal types of waste. Such countries opt for other disposal options such as landfill and recycling. Although this prohibition of incineration of municipal waste is unusual, when dealing with incineration upon the high seas, it is the rule. Specifically, the 1996 Protocol to the 1972 London Dumping Convention states 'Contracting Parties shall prohibit incineration at sea of wastes or other matter'.[25] In addition, the Parties must not allow the export of waste to other countries who seek to incinerate it upon the ocean.[26]

This conclusion with the 1996 Protocol followed 20 years of debate on the regulation of the highly contentious practice of incineration on the high seas under the auspices of the London Dumping Convention.[27] This debate, which began in 1976, was followed by the 1980 Technical

---

[24] Tyskeng, J. (2010). 'Comparing Energy Use and Environmental Impacts of Recycling and Waste Incineration'. *Journal of Environmental Engineering*. Aug. 742–53; Merrild, H. (2008). 'Life Cycle Assessment of Waste Paper Management: The Importance of Technology'. *Resources, Conservation and Recycling*. 52: 1391–8; Magrinho, A. (2008). 'Estimation of Residual MSW Heating Value as a Function'. *Waste Management*. 28: 2675–83; Schmidt, J. (2007). 'Life Cycle Assessment of the Waste Hierarchy: A Case Study on Waste Paper'. *Waste Management*. 27: 1519–30; Daskalopoulos, E. (2008). 'An Integrated Approach to Municipal Solid Waste Management'. *Resources, Conservation and Recycling*. 24: 33–50; Ekvall, L. (2000). 'Perspectives on Energy and Material Recovery from Paper'. *Transactions of the Institute of Chemical Engineers*. 78(B): 120–40.

[25] The 1996 Protocol to the Convention on the Prevention of Marine Pollution by Dumping and Other Matter, with Annexes (the 1996 Protocol). Article 5. According to article 1(5) of the 1996 Protocol to the London Dumping Convention, 'Incineration at sea' means the combustion on board a vessel, platform or other man-made structure at sea of wastes or other matter for the purpose of their deliberate disposal by thermal destruction. It does not include the incineration of wastes or other matter on board a vessel, platform or other man-made structure at sea if such wastes or other matter were generated during the normal operation of that vessel, platform or other man-made structure at sea.

[26] Article 6. 1996 Protocol.

[27] Anon. (1988). 'Incineration Ships Poison North Sea's Sediments'. *New Scientist*. 29 Sept. 33; Caufield, C. (1984). 'More Toxic Waste to be Burned at

Guidelines on the Control of Incineration of Wastes at Sea. These Guidelines set down requisite temperatures and efficiency requirements for combustion, site selection, retention of ash, and obligations of no discharge into the ocean of residue. Over the following decade, the standards in these Guidelines were progressively strengthened in terms of, inter alia, permitting of all applications; monitoring; flow, efficiency, minimum temperatures (1,200 °C) and destruction rates of the incineration process; site selection; and restriction of what wastes could, and could not, be incinerated. Compliance and liability regimes were also agreed.[28] The end result was that the standards became so high that incinerator ships ceased operating. From there, it was a small step to total prohibition of the incineration of waste on the high seas in 1996.

Whilst the industrial incineration of waste has been banned on the high seas, the same prohibition has not been applied on land. Here, nations, regions and the international community have opted to regulate, rather than prohibit, the practice. The basis of this regulation, whether it is in

---

Sea'. *New Scientist.* 12 Apr. 10; Anon. (1984). 'Sea Burn'. *New Scientist.* 3 May. 5; Anon. (1984). 'Toxic Burn Ban'. *New Scientist.* 7 June. 5.

[28] In terms of sequence on this matter, see Annex VI (LDC I, 1975). Resolution on Incineration at Sea; Annex II (LDC II, 1977). Incineration at Sea; Annex II (LDC III, 1978). Notification of Special Permits to Incinerate Wastes and Other Matters at Sea; Annex III (LDC III, 1978). Incineration at Sea; Annex IX (LDC IV, 1979). Preliminary Procedures for the Implementation of Paragraphs 8 and 9 of Annex I for the Purposes of Incineration at Sea; Annex VI (LDC V, 1980). Addition to the Interim Technical Guidelines on the Control of Incineration at Sea; Annex VII (LDC V, 1980). Export of Wastes for Incineration at Sea; Annex III (LDC 9, 1985). Interim Provisions for the Surveillance of Cleaning Operations Carried Out at Sea on Board Incineration Vessels; Annex IX (LDC 10, 1986). Proposed Resolution on New Guidelines for the Surveillance of Cleaning Operations Carried Out at Sea on Board Incineration Vessels; Annex VI (LDC 11, 1988). Guidelines for the Surveillance of Cleaning Operations [for] Incineration Vessels; Annex VIII (LDC 10, 1986). Terms of Reference for the Joint LDC/OSCOM Group of Experts on Incineration at Sea; Annex VII (LDC 11, 1988). Status of Incineration of Noxious Liquid Wastes at Sea; Annex V (LDC 11, 1988). Amendments to the Interim Technical Guidelines on the Control of Incineration of Wastes and Other Matters at Sea; Annex VI (LDC 12, 1989). Work Programme of the Scientific Group on Incineration at Sea, Annex III (LDC 13, 1990). Work Programme on Matters Related to Incineration at Sea; Annex II (LDC 13, 1990). Status of Incineration of Noxious Liquid Wastes at Sea; Annex V (LDC 15, 1992). Status of Incineration of Noxious Liquid Wastes at Sea; Annex IV (LDC 16, 1993). Amendments to Annex I Concerning Incineration at Sea.

Europe,[29] in the applicable guidelines of the Basel Convention[30] or in the 1991 Madrid Protocol on Environmental Protection to the Antarctica Treaty (the Madrid Protocol), is the same. Specifically, as the Madrid Protocol stipulated, the incineration of waste could only be undertaken 'in incinerators which to the maximum extent practicable reduce harmful emissions'.[31]

To achieve this goal of maximum reduction of harmful emissions, three principles have become clear. First, certain wastes should not be incinerated under any conditions. These include glass and heavy metals, and in particular, mercury. These are either pointless to incinerate or their hazardous nature will only end up being transferred to the ash or airborne emissions.[32]

Second, modern safety levels of pollutants from incinerators have to be met. These standards have become progressively stronger than their predecessors. For example, when the United States revised their Regulations Governing Emissions from Medical Waste Incinerators in 1997, the new rules aimed to reduce mercury emissions by 94 percent, particulate matter by 90 percent, hydrogen chloride by 98 percent and dioxin by 95 percent. Building on such heightened restrictions on pollutants from incinerators, organisations such as the World Health Organization have established a Provisional Tolerable Monthly Intake (but not emission limits) for dioxins, furans and polychlorinated biphenyls of 70 picograms ($10^{-12}$ g) per kilogram of body weight. These limits are an estimate of the amount of chemical per month that can be

---

[29] See articles 10, 23 and 38 and Annex II of Directive 2008/98/EC (19 Nov. 2008) on Waste.

[30] See the Basel Secretariat (1995). *Technical Guidelines on Incineration on Land* (Basel Secretariat, Basel, D10); also, Basel Secretariat (1999). *Technical Guidelines on the Environmentally Sound CoProcessing of Hazardous Wastes in Cement Kilns* (Basel Secretariat, Basel). For the adoption of these principles, see Decision V/26 (1999).

[31] Article 3 of Annex III on Waste Disposal. Madrid Protocol.

[32] Basel Convention (2002). *Technical Guidelines for the Identification and Environmentally Sound Management of Plastic Wastes and for their Disposal.* (Convention Secretariat, Basel). 29–31; UNEP (2004). *An Inventory of World-Wide PCB Destruction Capacity.* (UNEP, Geneva). ii–iii; Almedia, M. (2009). 'Laboratory Study on the Behaviour of Spent AA Household Alkaline Batteries in Incineration'. *Waste Management.* 29(1): 342–9; Coghlan, A. (1994). 'Can Burning Plastics be Good for the World?' *New Scientist.* 2 July. 20; OECD (2005). *Improving Recycling Markets.* (OECD, Paris). ENV/EPOC/WGWPR (2005)3/FINAL. 49–51; UNEP (1999). *Dioxin and Furan Inventories.* (UNEP, Geneva). 4–9.

ingested by a human over a lifetime without appreciable health risk. From such guidelines, nations or regions have gone on to set emission levels that may come from industrial incinerators of municipal, industrial, and/or specialist incinerators such as for medical waste. For example, Europe set specific emission levels from incineration plants for dioxin, furans, mercury, cadmium, thallium, arsenic, lead, chromium, copper, nickel and zinc.[33] These emission levels are commonly supplemented by standards for the operation of the plant which include the design, temperature, pollution traps and the geographical location of the site. All of these standards are a far cry from a few decades ago, with some factors, such as temperature from incineration, rising from 850 °C in the 1970s, to around 2,000 °C today.[34]

## 6. CONCLUSION

The burning of waste in the open is a practice which is uncommon in developed countries. This is not the trend in developing countries, where open burning of waste remains one of the most common methods to reduce the size of the waste pile and some of the nuisances associated with it. This ongoing practice is despite the clear environmental and social risks of poor incineration.

Although open burning is largely a historic practice in the developed world, the practice of controlled burning, via either municipal, industrialised or specialised incinerators, is a growth industry in both the

[33] See Annex IV, Emission limit values for discharges of waste water from the cleaning of exhaust gases, of Directive 2000/76/EC (4 Dec. 2000) on the Incineration of Waste. The rule is set out in Article 7. World Health Organization (2011). *Health-Care Waste Management* (WHO, Geneva, Fact Sheet No. 281). 3; Pavlas, M. (2011). 'Waste Incineration with Production of Clean and Reliable Energy'. *Clean Technology Environmental Policy.* 13: 595–605; Consonni, S. (2005). 'Alternative Strategies for Energy Recovery from Municipal Solid Waste'. *Waste Management.* 25: 123–48; McKenna, P. (2009). 'Into this Air'. *New Scientist.* 25 Apr. 33–5; Gelfand, L. (2001). 'Waste to Energy Incineration'. *Energy Engineering.* 98(1): 23–34; World Health Organization (1999). *Dioxins and their Effects on Human Health.* (WHO, Geneva. Fact sheet No. 225); Frosch, R. (1995). 'The Industrial Ecology of the 21st Century'. *Scientific American.* September. 144–8.

[34] World Health Organization (2011). *Health-Care Waste Management* (WHO, Geneva, Fact Sheet No. 281). 3. The European figures can be found in their Directive 2000/76/EC (4 Dec. 2000) on the Incineration of Waste. Article 6(3).

developed, and developing, world. This industry, which began in the 19th century, accelerated from the 1970s onwards when it became assumed that it was not only an effective and safe way to destroy waste, it was also one which could be turned to harnessing energy. However, the advocacy for the incineration of waste is troubled by a number of considerations. First, it is not always clear that incineration of waste (assuming perfect operations) is the best thing to do with it, in terms of life-cycle analysis. Second, just as with open burning, the evidence suggests that if controlled burning is done inefficiently, pollutants will result and these may cause both environmental and social damage. This is currently a particularly polarised debate, and in a contemporary setting, is very much determined by what is burnt, to what standard, and how exposed people are to the potential source.

Domestic, regional and international law have attempted to deal with these problems in a number of ways. On the high seas, the problem was dealt with by prohibition. On land, the imposition of high technical standards for the operation of incineration plants have been combined with emission limits for specific pollutants. These standards appear to have mitigated some of the impacts associated with earlier, more inefficient, forms of incineration. However, this appearance only exists for the countries which possess the high standards and the state of the art technology to achieve them. In most other places, the inefficient combustion of any type of waste, whether done in the open or in a controlled environment, represents an ongoing and direct environmental and social threat.

# 6. Disposal into the land

## 1. INTRODUCTION

This chapter is about the disposal of waste into (or onto) the land. The wastes that this chapter covers include all wastes except that which is nuclear. Nuclear waste, which is also placed into the land, is dealt with in Chapter 9. Placing waste into the ground is the most common form of waste disposal in both the developed and developing world.

Given that this is the dominant location for the disposal of waste, an accompanying large body of rules and principles has evolved to manage the associated problems in this area. These rules and principles have become well adopted and practised in most developed countries. The same cannot be said for many developing countries.

## 2. BEGINNINGS

Evidence of placing solid waste in an allocated spot on a map, and filling, tipping or dumping the waste into, or onto, the land began with the Cretan capital, Knossos, around 2000 BCE. These people of the early Minoan civilisation placed their solid waste in large pits and, as time progressed, covered the waste with layers of soil. By 500 BCE, this type of practice was being regulated, with ancient Athens being the first community to leave behind regulations detailing how the citizens organised the first municipal dump, in terms of pre-selected locations which had to be at least 1 mile from the city walls. The Romans followed this practice. In some instances, the size of the dumps, such as that at Monte Testaccio which held the waste of ancient Rome, reached a height of 165 feet, and 1,100 square yards. Two thousand years on, all that remains is perhaps 45 million broken earthenware storage jars.[1]

[1] Everitt, A. (2009). *Hadrian and the Triumph of Rome*. (Random House, NYC). 74–5; Franchetti, M. (2009). *Solid Waste Analysis and Minimisation*. (McGraw Hill, NYC). 23; Rathje, W. (1992). *Rubbish: The Archaeology of Garbage*. (HarperCollins, NYC). 19, 32–52.

Whilst it is likely that some of the rudimentary practices in landfill continued after the collapse of Rome, especially in those areas which remained as cities until the time of the Renaissance, such as within Byzantium, the topic of waste disposal on land did not clearly reappear until the 1300s. At this point, as populations started to grow and industries began to both evolve and expand in a number of European cities, such as tanners, potters and food supply, the need to control waste disposal became a much larger concern. Meat markets were a particular concern, especially after they had been linked to the plague of the Black Death. Accordingly, in 1388, the English Parliament proclaimed:

> For that so much dung and filth of the garbage and entrails as well of beasts killed, as of other corruptions, be … put in ditches, rivers, and other waters … within … cities, boroughs, and towns of the Realm … that the air there is greatly corrupt and infest … to the great annoyance, damage and peril of the inhabitants. It is accorded … that all they which do cast and lay all such annoyances, dung, garbage, entrails, and other ordure, in ditches, rivers, waters and other places … shall cause them utterly to be removed, avoided and carried away [or they will be fined].[2]

Despite such notations, for most of the Middle Ages the control of solid waste was a limited concern. However, when the industrial revolution began, and streams of modern municipal waste started to expand due to rising consumption and changes in type of items consumed, new understandings and political contexts began to emerge. Specifically, the management of waste became seen as a political goal, to help alleviate the plight of the industrial poor, and thus subvert possible revolutions. These concerns came as part of a wave of municipal involvement and active planning as the counterweight to curb some of the most obvious downsides of the industrial revolution, including, inter alia, the supply of clean water, urban planning, building codes, zoning, fire services, policing, courts, jails, harbours and piers, cemeteries, public hospitals, public schools – and the control of waste.[3]

---

[2] This 1388 'Legislation on Town Sanitation' from the English Parliament is reprinted in Baker, D. (ed.). *Portraits and Documents: England in the Later Middle Ages.* (Academica, London). 151.

[3] Pontin, B. (2012). 'Nuisance Law and the Industrial Revolution: A Reinterpretation of Doctrine and Institutional Competence'. *Modern Law Review.* 75(6): 1010–36; Rosenthal, L. (2007). 'Economic Efficiency, Nuisance and Sewage'. *Journal of Legal Studies.* 36(1): 27–62; Paavola, J. (2002). 'Water Quality as Property: Industrial Water Pollution and Common Law in the Nineteenth Century United States'. *Environment and History.* 8: 295–318;

The control of waste in modernity can be traced to 1757 when Benjamin Franklin, who considered cleanliness both a personal and social virtue, instituted the first municipal street cleaning service in Philadelphia. A few decades later in 1803, Washington, D.C. passed a law requiring that all filth or other offensive substances, nuisances or obstructions in a public space had to be removed by the person or company causing the problem. A further law in 1820 gave the mayor the authority to appoint scavengers for each ward who were to remove garbage and dead animals from the streets and to clean out privies if they were not adequately maintained. This was relatively easy to implement as human scavengers were as plentiful as solid waste – with over 10,000 dead horses per year in each of the big cities. The human scavengers were supplemented by tens of thousands of stray dogs, pigs and goats that ate their way through whatever edible waste they could find. These scavengers, human and animal, were replaced with organised, standardised collectors of rubbish and street cleaners. Over 70 percent of all American cities had such services by 1880. The exemplar of these was the 3,000 street cleaning employees of New York city, keeping the streets tidy of all forms of waste, in their immaculate white uniforms in the last decade of the century.[4]

Britain followed a similar pattern, with their first national law in this area, the 1848 Public Health Act, mandating the collection of solid waste in all cities. This law also required local authorities to cleanse the streets and remove dust and filth. This obligation was strengthened in 1875 in the amended Public Health Act, which obliged local authorities in Britain to, inter alia, not only collect all the rubbish in their municipalities, but

---

Brenner, J. (1974). 'Nuisance Law and the Industrial Revolution'. *Journal of Legal Studies*. 3: 403; McLaren, J. (1983). 'Nuisance Law and the Industrial Revolution: Some Lessons from Social History'. *Oxford Journal of Legal Studies*. 3: 155.

[4] Zimring, C. (2004). 'Dirty Work: Xenophobia and the American Waste Trades, 1870–1930'. *Environmental History*. 9: 80–97; Medina, M. (2001). 'Scavenging in America: Back to the Future?'. *Resources, Conservation and Recycling*. 31: 229–40; Crane, G. (2000). 'Filth, Garbage, and Rubbish: Refuse Disposal, Sanitary Reform, and Nineteenth-Century Yard Deposits in Washington'. *Historical Archeology*. 34(1): 20–38; Strasser, S. (1999). *Waste and Want: A Social History of Trash*. (Holt, NYC). 3–15, 29–31, 69–73; Rogers, H. (2005). *Gone Tomorrow: The Hidden Life of Garbage*. (Norton, NYC). 41–2, 83–5; Hoy, S. (1996). *Chasing Dirt: The American Pursuit of Cleanliness*. (Oxford, OUP). 14–15; Cumbler, J. (1995). 'Whatever Happened to Industrial Waste?'. *Journal of Social History*. 29(1): 149–71.

also ensure that, once it was collected, it was not wrongfully dumped in specific areas, such as waterways or village greens.[5]

## 3. EMERGING PROBLEMS

Whilst it was agreed early that there were some priority areas where municipal waste could not be dumped (such as in freshwater streams or valued common areas of the community), it was soon discovered that the sheer volume of what was now being collected and directed towards one concentrated point was creating a new generation of nuisance risks related to, inter alia, fire, odour, vermin and disease (especially when people and/or animals were/are in close proximity). To confront such first-generation problems of dumps, which continue to exist today in many developing countries in the 21st century, the developed world began to deal with them in the opening decades of the 20th century. Here, in the case of both Britain and the United States, it was recognised that although getting the waste out of the cities was a success, dealing with it in its final resting place on the land required much more thought.[6]

The answer to the problem emerged during the Second World War when the Army Corps of Engineers of the United States adopted the practice of sanitary landfill (where the waste was/is isolated from the environment) as a way to mitigate against the risks to the troops of the traditional nuisances associated with landfill, as their preferred method for disposal of waste for all American military facilities during the conflict. This practice spurred the American Society of Civil Engineers (in 1959) and the National Academy of Sciences (in 1966) to consolidate the principles of how to make landfills safe.[7]

---

[5] The 1848 Public Health Act. 11 and 12 Vict. c. 63. Article 55. As reprinted in Lumley, W. (1859). *The New Sanitary Laws*. (Shaw Printers, London). 1–222. The 1875 Act is reprinted in *Halsbury's Laws of England* (London, 3rd Edition, 1964). 26: 38.

[6] Guest, P. (2014). 'Mapping Nairobi's Disease Flashpoints'. *New Scientist*. 30 Aug. 14; Chemel, C. (2012). 'Odour-Impact Assessment Around a Landfill Site'. *Journal of Environmental Management*. 93: 85–94; Cook, S. (2008). 'Problem Bird Species on Landfill Sites'. *Environmental Management*. 41: 834–43; Sharp, T. (2007). 'On Rats, Refuse, and Recycling'. *Journal of Urban Health*. 84(5): 23–9.

[7] National Academy of Sciences (1966). *Waste Management and Control*. (NAS, Washington). Also, Colten, C. (1996). *The Road to Love Canal: Managing Industrial Waste Before the EPA*. (Austin, Texas UP). 25, 28–48, 58–9, 67, 86–7; Murphy, A. (2012). '"Litterers": How Objects of Physical Disorder Are

The problem was, and is, that the risks (and the understandings thereof) that landfills may pose have evolved. Thus, the first generation of problems related to the more immediate problems that landfills create, in terms of fire, odour and vermin, was relatively easy to understand. Even what may be considered third-generation problems, such as the contribution of landfill sites to global problems like climate change, are relatively easy to understand. That is, although each site is different, collectively, landfill sites, emitting methane due to degrading organic waste, are responsible for approximately 13 percent of the total methane emissions globally and up to 5 percent of the global greenhouse budget.[8] Where there is the greatest difficulty in understanding the negative impacts of landfill sites is with second-generation problems, such as leachate.

Leachate is a problem which first became identified in the 1930s. It occurs when contaminated effluent, most commonly liquid, emerges after having percolated through a waste site. Leachate may be virtually harmless or dangerously toxic, with at least 18 levels of classification risk being identified. These risks may appear over five decades after a site has closed. The ultimate outcome of the impact of any one of these depends upon two factors. The first factor is the site and the surrounding environment in which waste is dumped. The key considerations here are time, the moisture content of the site and its temperature, acidity, oxygen level and microbial activity. The broader geological context and the water flow through the site, in terms of groundwater, surface water and

---

Used to Construct Subjects'. *The ANNALS of the American Academy of Political and Social Science*. 642: 210; Paavola, J. (2006). 'Interstate Water Pollution Problems and Elusive Federal Water Pollution Policy in the United States, 1900–1948'. *Environment and History*. 12: 435–65; Hoy, S. (1996). *Chasing Dirt: The American Pursuit of Cleanliness*. (Oxford, OUP). 150–52; Rathje, W. (1992). *Rubbish: The Archeology of Garbage*. (Harper, NYC). 196.

[8] Habib, K. (2013). 'A Historical Perspective of Global Warming Potential from Municipal Solid Waste Management'. *Waste Management*. 32: 45–56; Zhang, C. (2012). 'Methane Distribution Surrounding Closed Landfill Sites'. *Environmental Technology*. 33(18): 2159–66; Chanton, J. (2011). 'Landfill Methane Oxidation Across Climate Types.' *Environmental Science and Technology*. 45: 313–19; IPCC (2007). *Fourth Assessment Report: Summary for Policy Makers*. (IPCC, Geneva). Paragraph 16; Themelis, M. (2007). 'Methane Generation in Landfills'. *Renewable Energy*. 32: 1243–57; Humer, M. (2006). 'Biotic Systems to Mitigate Landfill Methane Emissions'. *Waste Management Resources*. 26: 33.

precipitation, are critical, as the poisoning of associated freshwater systems is the most common impact of leachate.[9]

The second factor is the type of waste held in a site. Accumulated municipal waste with large amounts of organic material in it, even without conventional municipal waste added in, can produce leachate. However, when toxic materials are added, the problem magnifies, as the leachate and/or associated contamination produces a much higher risk of poison. This problem became very apparent in the 1970s and 1980s throughout the developed world as thousands of sites which had been used as dumping grounds for all types of waste became a very public problem. With the most high profile instance, Love Canal in New York State, it was found that the landfill, which began in the 1920s, held municipal, industrial and military waste, including that from the Manhattan Project, as well as 21,800 tons of chemicals that were dumped between 1947 and 1952. Twenty years later after the dump had been closed, had soil placed on top of it, and had been landscaped a community of over 1,000 people was built on top and around the old dumping ground. The problem was that although dumping had stopped decades earlier, 82 industrial chemicals were still evident in the air, soil and water, including 11 known or suspected carcinogens.[10]

---

[9] Tonjes, D. (2013). 'Classification Methodology for Landfill Leachates'. *Journal of Environmental Engineering*. 139(8): 119–30; Liehr, G. (2013). 'Assessing the Risk of a 50 Year Old Dump'. *Journal of Soils and Sediments*. 13(7): 1270–83; Gilbertson, K. (2012). 'Not a Load of Old Rubbish'. *The Safety & Health Practitioner*. 30(12): 44–6; White, R. (2011). 'Limitations in the Management of Hazardous Waste and Its Disposal'. *Journal of Environmental Protection*. 2(9): 1257–63; Singh, R. (2010). 'Review of Groundwater Contamination Hazard Rating Systems for Old Landfills'. *Waste Management and Research*. 28(2): 97–108; Koshi, E.L. (2007). 'Bioreactivity of Leachate from Municipal Solid Waste Landfills'. *Science of the Total Environment*. 384: 171–81; Li, M. (2012). 'Occurrence of PFCs and PBDEs in Landfill Leachates'. *Water, Air and Soil Pollution*. 223: 3365–72; Cumar, S. (2011). 'Environmental Impact of Leachate Characteristics on Water Quality'. *Environmental Monitoring and Assessment*. 178: 499–505; Wiszniowski, J. (2006). 'Landfill Leachate Treatment Methods: A Review'. *Environmental Chemistry Letters*. 4: 51–61; Richards, K. (1989). 'All Gas and Garbage'. *New Scientist*. 3 June. 20–24.

[10] Engelhaupt, E. (2008). 'Happy Birthday, Love Canal'. *Environmental Science and Technology*. 42(22): 8179–90; Colten, C. (1996). *The Road to Love Canal: Managing Industrial Waste Before the EPA*. (Austin, Texas UP). 6, 45–55, 66–7; Stigliani, W. (1991). 'Chemical Time Bombs'. *Environment*. 33(4): 5–9; Raloff, J. (1980). 'Disaster on 99th Street'. *New Scientist*. 19 June. 298; Pearce, F. (1992). 'Sitting on a Toxic Time Bomb'. *New Scientist*. 15 Aug. 12–13; Pearce, F. (1994). 'Gummer Buries List of Poisoned Land'. *New*

With many such toxic landfill sites, the evidence suggests that, depending on the two factors noted above (type of waste and surrounding environment), a detrimental impact upon local ecosystems and associated species is quite possible. Negative impacts have been shown with a number of non-human species that are both aquatic and terrestrial. Detrimental impacts have also been shown to human populations living in close proximity (broadly, between 1 and 3 kilometres) of a hazardous waste site. In this regard, small but significantly increased risks of developmental toxicity, generally in terms of low birth weights and birth abnormalities, occurred in babies whose mothers lived within 1 to 3 kilometres of a hazardous waste site. The development of certain types of cancers has been documented as well.[11]

The scientific problem with these types of findings is in clearly linking a particular pollutant from the specific site to the negative impact on a person who was near that location. The same difficulty exists in linking

---

*Scientist*. 3 Dec. 6; Anon. (1993). 'Foul Play over Toxic Dumps'. *New Scientist*. 6 Mar. 3; Pearce, F. (1992). 'Worst First on List of Contaminated Land'. *New Scientist*. 8 Aug. 5; Milne, R. (1990). 'Landfill Tips Going Unchecked'. *New Scientist*. 12 May. 6; Milne, R. (1986). 'Flash Point for Britain's Toxic Waste Cowboys'. *New Scientist*. 31 July. 20; Pearce, F. (1982). 'Toxic Waste: The Cowboys'. *New Scientist*. 1 Mar. 628; Pearce, F. (1982). 'The Wastelands of Britain'. *New Scientist*. 30 Sept. 888–94; Pearce, F. (1981). 'Toxic Dump: Tip of Iceberg?'. *New Scientist*. 8 Oct. 83; Caulfield, C. (1981). 'Lords Get Tough on Toxic Waste'. *New Scientist*. 10 Sept. 639; Caulfield, C. (1981). 'An Overburden of Toxic Waste'. *New Scientist*. 7 May. 344; Pearce, F. (1981). 'Toxic Waste: The Political Connections'. *New Scientist*. 14 May. 408.

[11] Currie, J. (2011). 'Superfund Cleanups and Infant Health'. *American Economic Review*. 101(3): 435–41; Li, H. (2010). 'Landfill Leachate Ingestion and Mouse Viscera'. *Journal of Hazardous Materials*. 174: 54–8; Ewence, A. (2010). 'Possible Developmental Toxicity Associated with Landfill Sites in the UK'. *Toxicology*. 278: 341–78; Vrijheil, M. (2009). 'Landfill Sites and Congenital Anomalies'. *Occupational and Environmental Medicine*. 66(2): 71; Deguchi, Y. (2007). 'Evaluation of Mutagenic Activities of Leachates'. *Mutation Research*. 627: 178–85; Gilbreath. S. (2006). 'Adverse Birth Outcomes Associated with Open Dumpsites'. *American Journal of Epidemiology*. 164(6): 158–69; Comba, P. (2006). 'Cancer Mortality in an Area Characterized by Multiple Toxic Dumping Sites'. *Annual New York Academy of Sciences*. 1076: 449–61; Orr, M. (2002). 'Elevated Birth Defects in Minority Children Living near Hazardous Waste Sites'. *International Journal of Hygiene and Environmental Health*. 205: 19–27; Anon. (2002). 'Birth Defects Linked to Landfill'. *The Ecologist*. 32(5): 10; Pukkala, E. (2001). 'Increased Incidence of Cancer and Asthma in Houses Built on a Former Dump'. *Environmental Health Perspectives*. 109(11): 1121–9; Vrijheil, M. (2000). 'Health Effects of Residence near Hazardous Waste Landfill Sites'. *Environmental Health Perspectives*. 108 (Supp): 101–12.

detrimental health impacts between incinerator plants and local populations, so too, with landfill sites, scientific problems of proof exist. The difficulty is that many of the studies in this area may involve incomplete cohorts, bias, lack of data on individual exposure, and most importantly, too many other explanations, especially because the dominant groups impacted upon – typically economically poor minorities – are exposed to multiple other sources of pollution which could also explain elevated rates of detrimental impacts. Thus, with one of the most well-studied toxic landfill sites in the world – Love Canal – no clear statistical differences could be found directly linking the site to a damaged local population next to it.[12] It is due to such problems that the international community has pledged itself to work harder in resolving questions of causation and the scale of impact between landfill sites and associated ecosystems and human communities.[13]

---

[12] Brender, N. (2011). 'Residential Proximity to Environmental Hazards and Adverse Health Outcomes'. *American Journal of Public Health*. 101(Supp): 12–20; Langlois, P. (2009). 'Maternal Residential Proximity to Waste Sites'. *Paediatric and Perinatal Epidemiology*. 23: 321–31; Vincenti, D. (2009). 'Risk of Congenital Anomalies Around a Municipal Solid Waste Incinerator'. *International Journal of Health Geographics*. 8: 1–10; Giusti, L. (2009). 'A Review of Waste Management Practices and Their Impact on Human Health'. *Waste Management*. 29: 2227–39; Suarez, A. (2007). 'Maternal Exposures to Hazardous Waste Sites'. AEP. 17(10): 772–7; Palmer, S. (2005). 'Risk of Congenital Anomalies after the Opening of Landfill Sites'. *Environmental Health Perspectives*. 113(10): 1362–70; Dummer, T. (2003). 'Prevalence of Adverse Pregnancy Outcomes Around Hazardous Industrial Sites'. *Paediatric and Perinatal Epidemiology*. 17: 250–55; Jarup, L. (2007). 'Down Syndrome in Births near Landfill Sites'. *Prenatal Diagnostics*. 27: 1191–6; Saffron, J. (2003). 'The Human Health Impact of Waste Management Practices'. *Management of Environmental Quality: An International Journal*. 14(2): 191–213; Gensburg, L. (2009). 'Mortality among Former Love Canal Residents'. *Environmental Health Perspectives*. 117(2): 179–200; New York State Department of Health (1978). *Love Canal, Public Health Time Bomb: A Special Report to the Governor and Legislature*. (Albany, NY: NYC Department of Health). 7–10.

[13] For the commitment, see Agenda 21 (1992). UNCED. Doc A/CONF.151/4. Chapter 20, paragraphs 20–23; for the work, see WHO (2007). *Population Health and Waste Management: Scientific Data and Policy Options*. (WHO, Geneva). iii.

## 4. EVOLVING STANDARDS

In the United States, the process of creating new standards to deal with the evolving problems of landfill began with the 1969 National Environmental Policy Act[14] (which began to use environmental impact assessments when planning landfill sites) and the 1970 Resource Recovery Act. These two pieces of legislation were primarily focused upon studies, research and investigations, although the 1970 Act also began to provide for the creation of guidelines for solid waste collection, transport, recovery and disposal.[15] Technical and financial assistance for States followed with the 1976 Resource Recovery Act.[16] Direct, federal intervention only occurred following the Love Canal incident. Specifically, in 1979, open dumping of municipal waste was expressly prohibited. Standards for dump sites became mandatory the following year when the Hazardous Waste and Consolidated Permit Regulations were passed. These regulations established the basic cradle-to-grave approach to hazardous waste management that exists today.[17] Subsequent legislation involved the 1984 Hazardous and Solid Waste Amendment,[18] which revised the criteria for landfill. This also required control and treatment of all contaminated surface water running off landfills and the phasing out of hazardous waste being dumped in landfill. Further additions on compliance in this area were added in 1992.[19]

In Britain, the modern development of clear laws to deal with the evolving problems of landfill pollution began with the 1972 Deposit of Poisonous Waste Act, which stipulated that no person could deposit poisonous waste on land if it was able to give rise to an environmental hazard.[20] This was followed with the 1974 Control of Pollution Act,

---

14 42 U.S.C. 4321.

15 Resource Recovery Act. 84 Stat. 1227 1970–1971. Also, Public Law 91-512-26 Oct. 1970. Section 101.

16 Resource Recovery Act. 90 Stat. 2795 1976. Also, Public Law 94-580-21 Oct. 1976. Section 1002(a)(4).

17 EPA. (1993). *Safer Disposal for Solid Waste: The Federal Regulations for Landfills*. (EPA, Washington). For the first rules on hazardous waste published in the Federal Register on 19 May 1980, see 45 FR 33066, 19 May 1980. Also, Tabatai, M. (2012). 'Comparing US and EU Hazardous Waste Liability Frameworks'. *Houston Journal of International Law*. 34: 653–80.

18 The Hazardous and Solid Waste Amendment. 98 Stat. 3221, 1984. Also, Public Law 98-616-8 Nov. 1984.

19 Federal Facility Compliance Act. 106 Stat. 1505, 1992. Also, Public Law 102-886-6 Oct. 1992.

20 Deposit of Poisonous Waste Act. Statute number 21 of 1972. Article 1.

which ensured that each disposal authority had to have adequate plans for waste disposal so as not to cause environmental hazards, and licensing regimes which dealt with questions of both quantity, and type, of wastes to be disposed of.[21]

Over the subsequent decades, such national laws in Europe became supplemented by regional standards such as the 1999 Directive on Landfill.[22] This Directive, which is supplemented by a collection of other directives on other waste streams, aims:

> [b]y way of stringent operational and technical requirements on the waste and landfills, to provide for measures, procedures and guidance to prevent or reduce as far as possible negative effects on the environment, in particular, the pollution of surface water, groundwater, soil and air, and on the global environment, including the greenhouse effect, as well as any resulting risk to human health, from landfilling of waste, during the whole-cycle of landfill.[23]

At the international level, the broad principles for making landfill safe were recorded in the Technical Guidelines on Specifically Engineered Landfill[24] of the Basel Convention on the Control of Transboundary Movements of Hazardous Waste and Their Disposal. These Guidelines supplement the soft international goals as found in Agenda 21 when the international community agreed 'to strengthen and increase national

---

[21] Control of Pollution Act. Statute number 40 of 1974. This regime was eventually replaced by Part II of the Environmental Protection Act. Statute number 25 of 1995.

[22] See Annexes I and II of Directive 1999/31/EC (26 April 1999) on the Landfill of Waste.

[23] Article 1. Within this regime, in Annex I, it was made clear that landfill sites be either for hazardous, non-hazardous or inert waste; the sites shall not be mixed. Each site must be permitted, after being registered in accordance with set criteria, in which operation, monitoring and control over all sites must be uppermost. In terms of general management, all classes of landfill have to take into account the physical environment (especially with regard to water), water control/drainage and leachate controls, with liners of various thickness depending on what type of waste is held, as well as mitigation techniques for nuisances and hazards (noise, odour, vermin, fire, and so on).

[24] Basel Convention (1995). *Technical Guidelines on Specially Engineered Landfill*. (Basel Convention Secretariat, Basel). These were adopted in 2006. See Decision VIII/17 (2006). Specially Engineered Landfills. Note also, Basel Secretariat (1994). *Basel Technical Guidelines on Waste Collected from Households*. (Basel Convention Secretariat, Basel).

waste … systems'. The developed world was to have these by the year 2000, and the developing, by 2010.[25]

The broad system that these exhortations have been reaching for can be divided into six principles.

The first principle, and perhaps the most fundamental of all principles in the area of trying to make standard landfill sites safer, is that certain wastes should never be deposited in them. The wastes that should not be deposited in standard municipal landfills are those which are hazardous. This is particularly obvious with three classes of pollutant. These are obsolete persistent organic pollutants,[26] mercury[27] and heavy metals such as, lead, iron, copper, zinc, nickel and cadmium.[28] Each one of these has the heightened capacity to have a detrimental impact on local populations and the surrounding, if not global, environment. Such types of hazardous waste need to be disposed of via methods such as, inter alia, chlorinated solvents, chemical oxidisation, thermal treatments, pyrolysis or high-tech incineration. If these options are not available, they should be deposited into the earth, if it can be done in specially engineered, highly secure

---

[25] Agenda 21 (1992). UNCED. Doc A/CONF.151/4. Chapter 21.17 and 21.18.

[26] GEF (2011). *Selection of Persistent Organic Pollutant Disposal Technology for the Global Environment Facility* (GEF, Washington). 3–4; Dvorska, A. (2012). 'Obsolete Pesticide Storage Sites and Their POP Release into the Environment'. *Environmental Science, Pollution Research.* 19: 1944–52; Weber, R. (2011). 'Persistent Organic Pollutants and Landfills – A Review of Past Experiences and Future Challenges'. *Waste Management and Research.* 29: 107; Eggen, T. (2010). 'Municipal Landfill Leachates: A Significant Source for Pollutants'. *Science of the Total Environment.* 408: 5147–57.

[27] Lindberg, S. (2005). 'Airborne Emissions of Mercury from Municipal Solid Waste'. *Journal of the Air & Waste Management Association.* 55(7): 859–69; Feng, X. (2004). 'Landfill is an Important Atmospheric Mercury Emission Source'. *Chinese Science Bulletin.* 49(19): 2068–702; Kim, K. (2002). 'Mercury Emissions as Landfill Gas Large-Scale Abandoned Landfill Sites'. *Atmospheric Environment.* 36(31): 4919–28.

[28] Chang, L. (2013). 'Soil Heavy Metal Pollution Assessment near the Largest Landfill of China'. *Soil and Sediment Contamination.* 22(4): 390–403; Tsydesnova, O. (2011). 'Chemical Hazards Associated with Treatment of Waste Electrical and Electronic Equipment'. *Waste Management.* 31: 45–58; Pin-Jing, H. (2006). 'In Situ Distributions and Characteristics of Heavy Metals in Full-Scale Landfill Layers'. *Waste Management Research.* 22: 3–23; Suna, A. (2005). 'Determination of Solid Waste Capacity for Selected Heavy Metals in Landfills'. *Journal of Hazardous Materials.* B121: 223–32; Baun, D. (2004). 'Speciation of Heavy Metals in Landfill Leachate: A Review'. *Waste Management and Research.* 22: 3–23.

storage settings. In this area, deep well injection – some 300 to 7,500 metres below the surface in sites of salt or limestone, completely isolated from water, biologically inactive, and impossible to leach – are used.[29]

Outside of the obvious hazardous waste that should never be deposited into landfill, sites should be divided into two classes for the taking of clean non-hazardous waste (as in, inert, non-hazardous waste like un-contaminated soil or concrete) and other types of non-hazardous waste. The difficulty with this theoretical division is that multiple parts of the modern waste stream, although not considered hazardous in isolation, if presented in larger quantities, would be. For example, old paint, domestic pesticides, pharmaceuticals, photographic chemicals, certain detergents, personal care products, waste oil, household products that contain small amounts of mercury, heavy metals or ozone depleting chemicals and electronic waste all contain materials that if multiplied have the capacity to cause much damage if they leach. For example, if nail polish, or even cigarette butts, were brought to a landfill site in 55-gallon drums, due to their cumulative hazard potential, they would not get permission to dump. The end result is that whether it is intentional or not, landfills often end up as sinks for hazardous materials.[30]

---

[29] Particular methods to note are alkali metal reduction, base-catalysed decomposition, catalytic hydrodechlorination, very high-tech incineration, gas-phase chemical reduction, photochemical dechlorination and catalytic dechlorination reaction, plasma arc, the potassium tert-butoxide method, subs and sub critical water oxidation, or thermal and metallurgical production of metals. For a full discussion and updated guidelines, see the Basel Convention (2012). *Technical Guidelines for the Environmentally Sound Management of Wastes Including Persistent Organic Pollutants.* (Basel Convention Secretariat, Basel). The Parties agreed in 2013 to update these to complement the Stockholm Convention on Persistent Organic Pollutants (POPs), see Decision BC-11/3 (2013). *Technical Guidelines for the Environmentally Sound Management of Wastes Consisting of, or Containing, Persistent Organic Pollutants.* For the safe disposal of mercury, see Basel Secretariat (2011). *Technical Guidelines for the Environmentally Sound Management of Wastes of, or Containing, Mercury.* (Basel Convention Secretariat, Basel). UNEP/CHW.10/6/Add 2/Rev.1. In 2013, the Parties agreed to update these Guidelines also. See Decision BC-11/5 (2013). Technical Guidelines for the Environmentally Sound Management of Wastes Containing or Contaminated with Mercury. And, Vanguilder, C. (2012). *Hazardous Waste Management.* (Mercury, Boston). 86–90, 98–103, 110–20.

[30] Xara, Z. (2013). 'The Leaching Potential of Spent Alkaline Batteries.' *Journal of Material Cycles and Waste Management.* 15(1): 61–72; Scharff, H. (2012). 'Landfills as Sinks for Hazardous Substances'. *Waste Management and Research.* 30(12): 1234–42; Ministry for the Environment (2012). *Consented Non-Levied Cleanfills and Landfills in New Zealand.* (MFE, Wellington). 5–7;

Due to the fact that the vast majority of landfill sites, even if holding prima facie non-hazardous waste, still present a risk to the local environment, a second principle is required. This is that the location of the land disposal sites must possess the correct geological and hydrological features, so that the site and the waste they hold are not susceptible to leaking leachate. Primarily, they should be isolated from water or associated habitats, such as being near coastal areas, wetlands, flood plains, as well as sources of ground water or underground water. In addition, the site should not be in an area of seismic activity or one that includes cavities, faults or sinkholes. Natural geological features for landfill sites should also be adopted that ensure their impermeable nature, such as having clay that is above half a metre thick.[31]

The third principle is that all pollutants must be stopped from leaving the site. This means for pollutants which are heavier than air and likely to leak, the site must be lined with an artificial skin. Requisite liners, in terms of type and/or thickness, are stipulated to different standards in accordance with the waste being stored, and whether the site is designed

---

Parascandola, J. (2012). *King of Poisons: A History of Arsenic*. (Potomac, Virginia). 14; Healton, C. (2011). 'Butt Really? The Environmental Impact of Cigarettes'. *Tobacco Control*. 17: 20–35; Lisbona, D. (2011). 'A Review of Hazards Associated with Lithium Batteries'. *Process Safety and Environmental Protection*. 89: 434–42; Whorton, J. (2009). *The Arsenic Century*. (OUP, Oxford). 134–9, 150–60, 303–15; Xara, Z. (2009). 'The Leaching Potential of Spent Alkaline Batteries'. *Waste Management*. 29(7): 2121–31; Slack, R. (2009). 'The Management of Household Hazardous Waste in the United Kingdom'. *Journal of Environmental Management*. 90: 36–42; Swati, M. (2008). 'Hazardous Organic Compounds in Urban Municipal Solid Waste'. *Journal of Hazardous Materials*. 160: 213–19; Fjelsted, L. (2007). 'Household Hazardous Waste'. *Waste Management and Research*. 25(6): 502–9; Slack, R. (2005). 'Household Hazardous Waste in Municipal Landfills'. *Science of the Total Environment*. 337(1): 119–37; Gronow, J. (2004). 'Hazardous Components of Household Waste'. *Critical Review in Environmental Science and Technology*. 34(5): 419–55; Strasser, S. (1999). *Waste and Want: A Social History of Trash*. (Holt, NYC). 26; Colten, C. (1996). *The Road to Love Canal: Managing Industrial Waste Before the EPA*. (Austin, Texas UP). 27–35; Rathje, W. (1992). *Rubbish: The Archaeology of Garbage*. (HarperCollins, NYC). 123–6; Kunreuther, H. (1991). 'Managing the Risks of Hazardous Waste'. *Environment*. 33(3): 12–15; O'Leary, P. (1988). 'Managing Solid Waste'. *Scientific American*. 259(6): 18–24.

[31] Laner, D. (2012). 'A Review of Approaches for the Long-Term Management of Municipal Solid Waste Landfills'. *Waste Management*. 32(3): 498–512; Raffaello, C. (2010). 'Technical Evolution of Landfilling'. *Waste Management*. 30(6): 947–8; Rathje, W. (1992). *Rubbish: The Archaeology of Garbage*. (HarperCollins, NYC). 87.

to hold wet or dry waste. The liner, which should have a guaranteed life of at least 50 years, creates an underground bladder that should prevent contamination of ground water by collecting leachate and channelling what often becomes tens of thousands of litres of water per day (depending on the size of the site) to nearby treatment facilities. Increasingly, double and sometimes even triple liner systems are being selected with liquid collection systems above, below and between liners.[32] As much as technology must be utilised to prevent leachate running from the bottom of the site, it must also be employed to stop pollutants that are airborne, such as volatile organic compounds and methane, also leaving the site. This landfill gas must be collected and stored, or if this is not possible, flared.[33]

The fourth set of requirements for a well-managed landfill is that the site needs to be carefully managed in terms of controlling access, active compaction and covering. Simple rules on design can help the minimisation of leachate, for example, by tipping vertically rather than horizontally. Compaction and then the prompt application of appropriate cover (which may include treated sludge in some instances) over the deposited waste, and then grading to encourage run-off rather than infiltration, may also be effective steps in preventing leachate.[34]

The fifth set of management considerations relate to dealing with the site once it has reached capacity. Thereafter, it needs to be carefully monitored, before being brought back into some types of productive social use, such as sports grounds, parks or industrial estates. In all instances, a comprehensive programme of monitoring is required to cover

---

32 Foo, K. (2009). 'An Overview of Landfill Leachate Treatment'. *Journal of Hazardous Materials*. 171: 54–60; Rogers, H. (2005). *Gone Tomorrow: The Hidden Life of Garbage*. (Norton, NYC). 18–19; Rathje, W. (1992). *Rubbish: The Archaeology of Garbage*. (HarperCollins, NYC). 87–8, 112–16; Hecht, J. (1998). 'What Rotten Luck'. *New Scientist*. 14 Mar. 16; Uehling, M. (1993). 'Keeping Rubbish Rotten to the Core'. *New Scientist*. 28 Aug. 12–13.

33 For the European rule on this, see article 4.2 of Annex II of Directive 1999/31/EC (26 April 1999) on the Landfill of Waste. For commentary on the technological possibilities, see Zhang, C. (2012). 'Methane Distribution Surrounding Closed Landfill Sites'. *Environmental Technology*. 33(18): 2159–66; Chanton, J. (2011). 'Landfill Methane Oxidation Across Climate Types.' *Environmental Science and Technology*. 45: 313–19; Themelis, M. (2007). 'Methane Generation in Landfills'. *Renewable Energy*. 32: 1243–57; Humer, M. (2006). 'Biotic Systems to Mitigate Landfill Methane Emissions'. *Waste Management Resources*. 26: 33.

34 UN-Habitat (2008). *Global Atlas of Excreta, Wastewater Sludge, and Biosolids Management*. (Habitat, Nairobi). 28.

inputs to, contents of and emissions from the site and the surrounding environment. It should be designed to provide advance warning of any unexpected problems and guidance on remediation. In addition, for decades after closure, the owner/operator must continue to maintain the necessary financial resources (such as via insurance) so as to be able to rectify any problems that may occur after it has closed.[35]

Finally, governments should clean up old landfill sites which are not contained and are causing detrimental impacts on the surrounding ecology and human populations. This obligation was softly recognised in 1992 in Agenda 21.[36] A number of governments have also accepted this obligation as a more forceful promise with regard to Antarctica, and a number of specific pollutants. With regard to the former, the signatories to the Madrid Protocol to the Antarctica Treaty promised 'past and present waste disposal sites on land and abandoned work sites of Antarctic activities shall be cleaned up by the generator of such wastes and the user of such sites'.[37] In terms of waste sites related to specific pollutants that are regulated under international law, the signatories to both the 2009 Stockholm Convention on Persistent Organic Pollutants (the Stockholm Convention)[38] and the 2013 Minamata Convention on Mercury have promised to endeavour to develop appropriate strategies for identifying, assessing and remediating contaminated waste sites which contain the pollutants that their Parties are obliged to control[39]

---

[35] UN Human Settlements Programme (2011). *Solid Waste Management in the World's Cities*. (UNSP, NYC). 108–10; Porter, R. (2002). *The Economics of Waste*. (Resources for the Future, Washington). 60–61; Bland, E. (2007). 'Garbage of Eden'. *New Scientist*. 14 Apr. 39–44.

[36] Agenda 21 called for all governments to 'establish and maintain inventories of hazardous wastes and their treatment/disposal sites, as well as of contaminated sites that require rehabilitation'. Chapter 20, paragraph 22(a). And 'Governments should promote identification and clean-up of sites of hazardous wastes in collaboration with industry and international organizations. Technologies, expertise and financing should be available for this purpose, as far as possible and when appropriate with the application of the "polluter pays" principle'. Agenda 21 (1992). UNCED. Doc A/CONF.151/4. Chapter 20, paragraph 22(g).

[37] See Annex (III), On Waste Disposal and Waste Management, of the 1991 Madrid Protocol to the Antarctica Treaty, see articles 1, 2, 8, 9 and 10. Also, Stark, J. (2006). 'Abandoned Antarctic Waste Disposal Sites: Monitoring Remediation Outcomes and Limitations at Casey Station'. *Ecological Management and Restoration*. 7(1): 21–8.

[38] Article 6(1)(e). 2009 Stockholm Convention on Persistent Organic Pollutants.

[39] Article 12. 2013 Minamata Convention on Mercury.

Such obligations to clean up contaminated sites can also be found at the domestic level in a number of developed countries. The foremost example in this area is the United States with their Comprehensive Environmental Response, Compensation and Liability Act (or 'Superfund') of 1980[40] and 1986.[41] Under this legislation, tens of thousands of old landfill sites, accompanied by other contaminated sites, were submitted for inclusion on the National Priority List (which evaluates sites in relation to their toxicity). As of 2013, 1,627 were judged toxic enough to be on the list. In Europe, the list may be considerably bigger (although the threshold for qualification to the list may be lower), with possibly hundreds of thousands of sites being recognised as contaminated (of which 38 percent are former municipal and/or industrial waste landfills). The costs so far to remediate the sites on the list have been in the region of 30 billion USD, of which the majority has been sourced, after great legal and policy contention, from those who were originally responsible for the dumps. In the case of the United States, so far, 343 sites have been remedied and removed from the list after successful remediation, with each one taking, on average, about ten years of work. In Europe, the estimated costs for the annual management of contaminated sites is put at around 6 billion euros.[42]

The fact that a number of developed countries are starting to make progress in this area of remediation of old contaminated sites is reflective of the wider trend that most developed countries have made progress in setting robust principles, broadly in line with the six areas set out above, for contemporary landfill sites and dealing with the older and problematic sites. For a country such as the United States where over 60 percent

---

40 P.L. 96-510.

41 The Superfund Amendments and Reauthorization Act of 1986. P.L. 99-499.

42 Panagos, P. (2013). 'Contaminated Sites in Europe'. *Journal of Environmental and Public Health.* 18: 1–11; EPA (2011). *The Office of Solid Waste and Emergency Response: Fiscal Year 2010 End of the Year Report.* (EPA, Washington). 7–8; United States Government Accountability Office. (1999). *Superfund: Progress, Problems and Outlook.* (UNGAO, Washington). 7–10; Curie, J. (2011). 'Superfund Cleanups and Infant Health'. *American Economic Review: Papers & Proceedings.* 101: 3, 435–41; Barnett, E. (1994). *Toxic Debts and the Superfund Dilemma.* (Chapel Hill, NYC); Porter, R. (2002). *The Economics of Waste.* (Resources for the Future, Washington). 219–20, 234; Russel, M. (1992). 'The US Hazardous Waste Legacy'. *New Scientist.* July. 12–15. Note, not all old sites are wastelands. Some, when contained, can go on to become sites where certain types of biodiversity can rebuild and thrive. See Pearce, F. (2013). 'Living Treasures of Brownfield Badlands'. *New Scientist.* 13 July. 16.

of their waste is landfilled, this is a progressive sign. In all such instances in the developed world, although considerable work remains to be done, at least policy makers know what needs to be aimed for.[43]

The same progress cannot be said for much of the developing world. Broad figures suggest that there are tens of thousands of poorly regulated, existing, open non-sanitary sites, and problematic contaminated sites which already exist and are not being adequately addressed. In many of these countries there is often variable and inadequate levels of service, lack of environmental control systems, indiscriminate dumping, scavenging and poor environmental and waste awareness and standards for landfill. As an overall statistic, somewhere between 50 percent and 70 percent of the waste landfilled in the developing world is done so in an unsafe manner. In China, where over 91 percent of their waste is landfilled, despite possessing comprehensive standards, most of their landfills are more or less open dumps. Similarly, in India, about 90 percent of municipal waste is disposed of in open and unregulated landfills.[44]

---

[43] The OECD has recommended that all of its members should have adequate regulatory and enforcement infrastructure at appropriate governmental levels, ensuring that waste management facilities utilise the best available technologies, utilise appropriate technical guidelines (such as those from the Basel Convention) and have environmental liability regimes for those carrying out potentially risky activities. See Recommendation 1. OECD (2007). *Guidance Manual on Environmentally Sound Management of Waste.* (OECD, Paris). 22, 25, 33, 35. Also, EPA (2011). *Municipal Solid Waste Generation, Recycling, and Disposal in the United States: Facts and Figures.* (EPA, Washington, EPA-530-F-11-005). 3–4.

[44] World Bank (2012). *What a Waste: A Global Review of Solid Waste Management.* (World Bank, Urban Development Series, No. 15, Washington). 5, 13–17, 22, 26; UNEP (2012). *Global Environment Outlook 5.* (Progress Publishers, Malta). 173; Blacksmith Institute (2012). *Top Ten Toxic Pollution Problems: Landfill.* (Blacksmith, London). 3; UN Human Settlements Programme (2011). *Solid Waste Management in the World's Cities.* (UNSP, NYC). 15, 24; Zhang, D. (2010). 'Municipal Solid Waste Management in China: Status, Problems and Challenges'. *Journal of Environmental Management.* 91(8): 1623–33; Barley, S. (2009). 'Illegal Toxic Waste Spotted From Space'. *New Scientist.* 3 Oct. 9; Giusti, L. (2009). 'Waste Management Practices and Their Impact on Human Health'. *Waste Management.* 29: 2227–39; Kumar, A. (2009). 'An Assessment of the Status of Municipal Solid Waste Management in India'. *Waste Management.* 29: 883–95; Sharholy, M. (2008). 'Municipal Solid Waste Management in Indian Cities: A Review'. *Waste Management.* 28(2): 459–67; Matete, N. (2008). 'Towards Zero Waste in Emerging Countries'. *Waste Management.* 28: 1480–92; Lakshmikantha, H. (2006). 'Report on Waste Dump Sites Around Bangalore'.

## 5. CONCLUSION

Humanity began disposing of waste into the earth around 2000 BCE. This practice has continued for four thousand years. The difference today is that the amounts deposited are infinitely greater, as are the types and the locations it may derive from, for in the 21st century, the trade in all types of waste is a truly global phenomenon. To mitigate against these risks, since the 19th century, successive generations have developed rules and policies to control and regulate the disposal of waste into the land as the environmental and social consequences of waste which has been poorly disposed of on, or in, land has become apparent. The essence of these rules is in restricting the types of waste that can be disposed of into the land; the location of the site; mechanisms for capturing and dealing with all of the pollutants; and comprehensive regimes to deal with the sites when closed and/or in need of restoration due to earlier contamination. These rules have been largely adopted across most of the developed world. The same cannot be said for the developing world, where the practice of unregulated landfill appears to be growing with little restraint.

---

*Waste Management*. 26(6): 640–50; Tolba, M. (1993). *The World Environment.* (Chapman, London). 265; World Commission on Environment and Development (1987). *Our Common Future*. (OUP, Oxford). 227.

# 7. Disposal into the ocean

## 1. INTRODUCTION

When it was impractical to incinerate or dump waste into the land, the ocean had always been the second best target. Whilst domestic restraints in this area began in the early part of the 20th century, international ones did not emerge until the end of the 20th century. The international restraints, as found through the 1972 Convention on the Prevention of Marine Pollution by Dumping of Waste and Other Matter (the London Dumping Convention), were the subject of intense debate for over a decade, as the regime prohibited the dumping of only the most hazardous of substances. This approach was completely turned around in 1996 when a Protocol to that Convention reversed the assumption that it was safe to dump anything into the ocean. Now, only a few select waste streams may be deposited into the seas, and even then, under strict conditions.

## 2. BEGINNINGS

Traditionally, where waste was not burnt or dumped into the land it was disposed of into the ocean or rivers. When the stream of waste was small and relatively inert, it was a widely held belief that a few cartloads off a bridge, or a few bargeloads tipped into the ocean, would quickly be dissipated by the power and regenerative forces of the waterways. This was only considered to be a problem when waste was dumped too close to shore, or in the wrong tides, and tons of clotted, rotting flotsam would foul beaches, contaminate drinking sources or congregate around city piers. Recognition that this was a problem emerged in Europe and the United States in the later part of the 19th century, from which legislation quickly followed.[1]

---

[1] Rogers, H. (2005). *Gone Tomorrow: The Hidden Life of Garbage.* (Norton, NYC). 37–9, 68–9; Swanson, R. (2004). 'Science, Policy and the Management of Sewage Materials: The New York Experience. *Marine Pollution Bulletin.* 49(9):

For example, the 1889 Public Works on Rivers and Harbors Act of the United States declared it was, prima facie, not lawful:

> To throw, discharge, or deposit, or cause, suffer, or procure to be thrown, discharge, or deposit either from or out of any ship, barge, or other floating craft of any kind, or from the shore, wharf, manufacturing establishment, or mill of any kind, any refuse matter of any kind or description whatever other than that flowing from streets and sewers and passing therefrom in a liquid state, into any navigable water of the United States.[2]

Despite the apparent comprehensive nature of this Act, the reality was that it was only concerned with the waste stopping the navigation of waterways. It was not concerned with the environmental and social implications of waste being placed into the ocean. This type of concern evolved in the early part of the 20th century as legislators concerned with the dumping of waste into the ocean began to restrict dumping zones of various types of waste to particular locations. These locations were increasingly pushed further out to sea and into ocean currents that swept the waste away from the land.[3] Whilst this approach worked for limited amounts of non-hazardous domestic waste, as the scale of hazardous waste started to rapidly increase in the middle of the 20th century (with over 1 million tons of hazardous waste – mostly dredging – going into the ocean each year in the 1950s), the wisdom of such an unregulated approach began to be disputed.[4]

The suitability of the ocean as a dump for waste became particularly obvious with the disposal of high-level nuclear waste into the oceans, as unlike many of the other types of waste being disposed of into the ocean at this point, high-level nuclear waste was clearly immediately toxic and global in impact. The initial response of the international community agreed to in the 1958 Geneva Convention on the High Seas was for all countries to 'take measures to prevent pollution of the seas from the dumping of radioactive waste, taking into account any standards and regulations which may be formulated by the competent international

---

679–90; Strasser, S. (1999). *Waste and Want: A Social History of Trash*. (Holt, NYC). 126–8.

[2] Section 13. 1899 Public Works on Rivers and Harbors Act. 30 Stat 1121 (1896–1899). Also, Benidickson, J. (2007). *The Culture of Flushing: A Social and Legal History of Sewage*. (UBC Press, Vancouver). 48–51.

[3] Franchetti, M. (2009). *Solid Waste Analysis and Minimisation*. (McGraw Hill, NYC). 25; Rogers, H. (2005). *Gone Tomorrow: The Hidden Life of Garbage*. (Norton, NYC). 69, 82–3.

[4] Tolba, M. (1993). *The World Environment*. (Chapman, London). 265.

organizations'.[5] The only problem was that, although the practice developed where high-level waste was no longer dumped, medium- and low-level waste continued to be discharged into the oceans by multiple countries. Accordingly, it became apparent in the late 1960s that the goal of protecting the world's oceans from all types of nuclear waste would best be achieved by an international organisation specifically dedicated to this task.[6]

This goal of controlling the dumping of nuclear waste into the oceans was then joined with the general dumping of all forms of hazardous waste into the ocean at the 1972 Stockholm Conference on the Human Environment. The countries which attended this gathering then went on to call for states to 'take all possible steps to prevent pollution of the seas by substances that are liable to create hazards …'.[7] In addition, it was agreed that each country would:

> Ensure that ocean dumping by their nationals anywhere, or by any person in areas under their jurisdiction, is controlled and that Governments shall continue to work towards the completion of, and bringing into force as soon as possible of, an over-all instrument for the control of ocean dumping.[8]

## 3. THE LONDON DUMPING CONVENTION

Very soon after the Stockholm Conference, the 1972 Convention on the Prevention of Marine Pollution by Dumping of Waste and Other Matter was concluded.[9] Four years after the London Dumping Convention entered into force, the United Nations Convention on the Law of the Sea (UNCLOS) was concluded. The UNCLOS carried through a large

---

[5] Article 25. 1958 Convention.

[6] Alley, W. (2011). *Too Hot to Touch: The Problem of High Level Nuclear Waste*. (Cambridge UP, Cambridge). 40–41; Hamblin, J. (2009). *Poison in the Well: Radioactive Waste in the Oceans*. 153–75; Council on Environmental Quality (1970). *Ocean Dumping: A National Policy – A Report to the President*. (CEQ, Washington). 7–13; Schenker, M. (1973). 'Saving a Dying Sea? The London Convention on Dumping'. *Cornell International Law Journal*. 7: 32–47.

[7] Principle 7 of the 1972 Stockholm Declaration of the United Nations Conference on the Human Environment. UN Doc A/CONF.48/14. 1972.

[8] Recommendation 86(c) of the Plan of Action from Stockholm Conference, ibid, Annex II.

[9] The Convention is reprinted in Kiss, A. (1980). *Selected Multilateral Treaties in the Field of the Environment*. (UNEP, Nairobi). 283.

number of the obligations of the London Dumping Convention, making them applicable to all countries – including those which had not signed the 1972 Convention.[10] In addition, under the UNCLOS, all states were obliged to 'protect and preserve the marine environment, via minimiz-[ing] to the fullest possible extent the release of toxic, harmful or noxious substances, especially those which are persistent, from … dumping'.[11] It was added that any national standards in this area could not reflect levels that were 'less effective in preventing, reducing and controlling such pollution than the global rules and standards'.[12] The end result is through this link with the UNCLOS, that is, the 1972 London Dumping Convention is the premier instrument in international law dealing with waste disposal into the ocean.

The context of the 1972 London Dumping Convention was the realisation that the marine environment was of 'vital importance to humanity' and that 'the capacity of the sea to assimilate wastes and render them harmless, and … to regenerate … is not unlimited'.[13] Accordingly, the Parties promised to work cooperatively 'to promote the effective control of all sources of pollution of the marine environment' and 'take all practicable steps to prevent the pollution of the sea by the dumping of waste and other matter that is liable to create hazards to human health … marine life [and other] amenities'.[14]

To achieve this goal, the London Dumping Convention established a system of the regulation of ocean dumping built around monitoring,[15]

---

[10] Article 210(4). UNCLOS.

[11] See articles 192 and 193 of the UNCLOS. Annex II (LDC 17, 1994). Requirements of the UNCLOS with Particular Reference to Waste Disposal at Sea. For commentary, see Hoffmeyer, M. (1985). 'Ocean Dumping Provisions of the Convention on the Law of the Sea'. *Brooklyn Journal of International Law*. XI: 350–77.

[12] Article 210(6). UNCLOS. Note also article 197. The importance of monitoring, scientific information and reporting were also set down. See articles 204, 205 and 206.

[13] Preamble. 1972 London Dumping Convention.

[14] Articles 1 and 7. 1972 London Dumping Convention.

[15] Article 6(d). 1972 London Dumping Convention. For the rules on monitoring, see Annex VII (LDC IV, 1979). Form of Report for the Acquisition of Data on Monitoring of Dumping Sites; Annex VII (LDC 10, 1986). Reporting of Monitoring Activities; Annex II (LDC 12, 1989). Monitoring Activities; Report of the Fifth Meeting of the Consultative Parties. (IMCO, London). 8–9.

reporting[16] and scientific research.[17] This system was based upon three tiers (Annexes I, II and III) of waste categorisation, with each categorisation bringing with it different regulatory requirements.[18]

Annex I of the London Dumping Convention contained what was known as 'the Black List'. Dumping of anything inscribed on this list was prohibited, unless it was an emergency.[19] If something was not listed in Annex I, then a party could not be prevented from dumping it into the ocean.[20] To be placed on the Black List, the wastes had to possess 'intrinsic properties' that were 'simultaneously toxic, persistent and bio-accumulative', and would thus represent a strong potential harm to either humanity or the marine environment. In addition, it had to have an impact which was more than fleeting in the marine environment. If the ocean allowed the waste which was dumped into it to be 'rendered rapidly harmless' and leave only 'trace contaminants' which could not 'cause acute or chronic toxic effects or bioaccumulation in sensitive marine organisms' then it could not be listed in Annex I. Unsurprisingly,

---

[16] The reporting obligation can be found in article 6 of the 1972 London Dumping Convention. The resolutions on the notification, and sharing of information procedures, evolved over a small period of time. The process can be traced through Annex IV (LDC I, 1976). Draft Interim Procedure on Notification under Article VI(4) of the Convention; Annex VI (LDC IV, 1979). Report of the Disposal of Wastes Carried Out at Sea; Annex II (LDC VI, 1981). Procedure for the Notification of Permits Issued for the Dumping of Wastes and Other Matter at Sea; and Annex III (LDC 15, 1992). Procedure for the Priority Consideration of Amendments to the Convention. Report of the 20th Consultative Meeting. LC 20/14 (1999). Paragraph 2.6; Report of the 21st Consultative Meeting. LC21/13 (1999). Paragraph 4.2.

[17] Note, these points can also be traced to the 1972 Stockholm Conference, *supra* note 8. See Recommendations 87, 90, 91 and 92.

[18] See article 15 of the 1972 London Dumping Convention. Also, Annex IV (LDC IV, 1979). Terms of Reference for the Ad-Hoc Scientific Group on Dumping; Annex II (LDC V, 1980). Procedures for the Circulation of Proposed Amendments to the London Dumping Convention; Annex IV (LDC V, 1980). Procedure and Method of Approach for Preparing and Maintaining a List of Hazardous Substances or Groups of Substances; Annex III (LDC 8, 1984). Terms of Reference for the Scientific Group on Dumping; Annex VI (LDC 10, 1986). Future Work Programme of the Scientific Group on Dumping; and Annex V (LDC 21, 1999). Terms of Reference of the Scientific Group.

[19] Article 4. 1972 London Dumping Convention. For the emergency situation, see article 4(4)(e). Also, Annex V (LDC V, 1980). Procedures and Criteria for Determining Emergency Situations. Note the earlier Annex III (LDC I, 1976) on the same point.

[20] Article 4(3). 1972 London Dumping Convention.

the Parties then went to work for the next 20 years trying to define each of these key terms.[21]

In practice, Annex I, with its three interlocking negative intrinsic properties quickly came to encompass the most obvious hazardous pollutants about which there was little debate, such as mercury, cadmium, the most toxic pesticides, plastics, crude oil, chemical or biological weapons and high-level radioactive waste. However, wastes which could not satisfy all three criteria during the 1970s and 1980s, such as some types of heavy metals like lead, could not be placed on the Black List.

Annex II of the London Dumping Convention contained what was known as 'the Grey List'. These were wastes that could be dumped via a special permit, but care was needed when doing so.[22] The identified list covered a number of heavy metals and poisons such as arsenic, lead, copper, zinc, cyanides, pesticides (those not covered by Annex I), chromium, nickel, scrap metal and bulky wastes likely to sink to the sea bottom and 'radio-active wastes or other radio-active matter not included in Annex I' (such as medium- and low-level nuclear waste).[23] The threshold for entering into Annex II was easier than Annex I, for whilst Annex I substances were meant to be toxic, persistent *and* bio-accumulative, Annex II substances had to only trigger one of the three concerns. However, in practice, even triggering just one of these concerns for a waste stream which was not immediately and obviously hazardous proved very difficult.[24]

---

21 For the decisions on Annex I, see the Decisions from the Parties, namely, Annex II (LDC IV, 1979). General Guidelines for Classification of Substances to Annexes I and II of the LDC; Annex VIII (LDC V, 1980). Amendment of the Lists of Substances Contained in Annexes I and II; Annex II (LDC 9, 1985). Criteria for the Allocation of Substances to the Annexes; Annex III (LDC 11, 1988). Amendments to the Guidelines for Allocation of Substances to the Annexes. For the exceptions see Annex VI (LDC III, 1978). Interim Guidelines for the Implementation of Paragraphs 8 and 9 of the LDC; Annex V (LDC IV, 1979). Interim Guidelines for the Implementation of Paragraphs 8 and 9 of the LDC; Annex III (LDC 10, 1986). Guidelines for the Implementation of Paragraphs 8 and 9 of Annex I.

22 Article 4(b). 1972 London Dumping Convention.

23 Annex II. Paragraph c.

24 Annex II (LDC IV, 1979). General Guidelines for Classification of Substances to Annexes I and II of the LDC; Annex VIII (LDC V, 1980). Amendment of the Lists of Substances Contained in Annexes I and II; and Annex IV (LDC 10, 1986). Amendment to the Lists of Substances Contained in Annex

Annex III dealt with the 'White List'. Wastes on this list were meant to be inert and/or of natural origin and thus represent little threat to the ocean. As such, they could be dumped with only a general permit. The White List ended up being not so much a list of wastes, but of 17 characteristics of the waste and the proposed dump site that could exclude it from being on the White List, and thus justify the waste going on the Grey or Black List. The nine characteristics about the waste (such as amount, form, properties, toxicity, persistence, accumulation – ability to change in, and impact upon the environment) were matched with eight considerations on the dump site (ranging from location to dispersal rates).[25]

## 4. THE PROTOCOL TO THE LONDON DUMPING CONVENTION

Despite the fact that the London Dumping Convention grew rapidly in membership over the 1980s and that many of its core goals were made universal by the UNCLOS, by the end of that decade, it was apparent to most that it was a convention which was, fundamentally, out of date in both its thinking and orientation. Progress on restricting waste dumping into the ocean was painfully slow, with certain issues like nuclear dumping, as discussed in Chapter 9, consuming decades of acrimonious debate. This problem was reflected at all levels of the Convention as the Annexes, which were in essence the engine room of the Convention, were under a constant state of review but were only rarely amended due to the need for a two-thirds majority of the Parties to agree. Moreover, to make matters even worse, as the standards for dumping waste into the ocean became progressively lower than those for dumping of waste on land, countries began to ship their waste to other countries, knowing that the importer would dump it at sea.[26] Finally, the Convention began to

II. Also, Hughes, E. (1988). 'Ocean Dumping and its Regulation in Canada'. *Canadian Yearbook of International Law*. 26: 155–74.

[25] For the Decisions in this area, see Annex II (LCD VIII, 1984). Guidance on the Application of Annex III; Annex V (LDC 10, 1986). Amendments to Annex III, Annex IV (LDC 11, 1988). Guidelines for the Implementation and Uniform Interpretation of Annex III; and Annex III (LDC 12, 1989). Amendment to Annex III.

[26] Note, this practice was later prohibited under article 6 of the 1996 Protocol to the London Dumping Convention. An exception to this was agreed in 2009, with the export of carbon dioxide streams or disposal, in accordance with Annex

become moribund, as countries moved towards finding regional, as opposed to international, solutions to their local problems of dumping waste into the ocean. This was especially the case with nuclear waste disposal into the ocean.[27]

Due to such concerns, the Parties to the London Dumping Convention agreed that there would have to be an amendment conference which would allow an 'overall and thorough review' of the London Dumping Convention.[28] The international community supported this review, calling

---

I (based upon mutual consent, and minimal standards). See Annex V. Amendment to Article 6 of the London Protocol. (LCD 31/ LP 3, 2009). For the solidification of this position under the London Dumping Convention, see Annex VII (LDC 12, 1989). Resolution on the Export of Wastes for Disposal at Sea; Annex XIII (LDC 10, 1986). Export of Wastes for Disposal at Sea. Note also Annex III of the 14th meeting of the Parties to the London Dumping Convention of 1991; Annex IV. Control of Transboundary Movements of Waste for Disposal at Sea. Finally, Annex VIII (LDC 13, 1990). Matters Related to the Basel Convention on the Transboundary Movements of Hazardous Wastes and Their Disposal; Nollkaemper, A. (1991). 'Transboundary Movement of Hazardous Wastes for the Purpose of Dumping at Sea'. *Marine Pollution Bulletin.* 22(8): 377–80.

[27] See, for example, Skjaerseth, J. (2006). 'Protecting the North East Atlantic: Enhancing Synergies by Institutional Interplay'. *Marine Policy*. 30(2): 17–70. Also, Coenen, R. (1997). 'Dumping of Wastes at Sea: The Adoption of the 1996 Protocol to the London Convention'. *Review of European Community and International Environmental Law*. 6(1): 54–61. Note, article 8 of the 1972 London Dumping Convention and article 12 of the 1996 Protocol – both of which assert the importance of synergies in this area. For the much more assertive regional conventions on the nuclear waste question, the 1985 South Pacific Nuclear Free Zone treaty enjoined its contracting Parties, in Article 7, 'not to dump radioactive waste and other radioactive matter at sea anywhere within the South Pacific Nuclear Free Zone…'. This treaty is complemented by the Regional Convention for the Protection of the Natural Resources and Environment of the South Pacific Region where Parties 'agree to prohibit the dumping… and the disposal into the seabed and subsoil of radioactive wastes and other matter in the Convention area' (Article 10). The Pacific treaties were building on earlier European examples, such as the Protocol for the Prevention of Pollution of the Mediterranean Sea by Dumping from Ships and Aircraft (Barcelona, 1976) of the Convention of the Mediterranean Sea against Pollution which prohibited the dumping into the Mediterranean Sea of 'High- and medium- and low-level radioactive waste or other high- and medium- and low-level radioactive matter'.

[28] Annex IV (LDC 17, 1994). Amendments and Amendment Proposals; Annex II (LDC 16, 1993). Review of the London Convention; Annex IX (LDC 14, 1991). Resolution for the Convening of an Amendment Conference of the LDC; Annex V (LDC 14, 1991). Statement of the Fourteenth Consultative

for the Convention to be remodelled, to allow the taking of 'appropriate steps to stop ocean dumping … of hazardous substances'.[29] The review concluded in the 1996 Protocol to the 1972 Convention on the Prevention of Marine Pollution by Dumping of Wastes and Other Matter (the 1996 Protocol).[30] This Protocol came into force in 2006.[31]

The 1996 Protocol set itself the objective to 'prevent, reduce and where practicable eliminate pollution caused by dumping'.[32] To do this, the Parties agreed to take a wider view of what constituted dumping and the area covered (thus clearly covering the seabed as well as the ocean).[33] They also agreed that compliance was an issue that had to be taken seriously,[34] along with assistance to developing countries.[35]

---

Meeting of the Parties; Annex IV (LDC 13, 1990). Statement by the Thirteenth Meeting of the Parties to the LCD; and Annex IX (LDC 12, 1989). A Long Term Strategy for the London Dumping Convention.

[29] Agenda 21 (1992). UNCED. Doc A/CONF.151/4. Chapter 17.30(b).

[30] 36 ILM (1997). 1.

[31] Article 25. 1996 Protocol. It was envisaged that the Protocol would supersede the Convention as between the contracting Parties to this Protocol that are also Parties to the Convention, for which a transition period was factored in (see articles 23 and 26), along with guidance by the Parties themselves as the transition took longer than originally envisaged. See Annex III (LDC 20, 1999). Conclusions of the Ad-Hoc Group on the Arrangements and Legal Issues Concerning the Relationship between the 1972 London Dumping Convention and the 1996 Protocol.

[32] Article 2. 1996 Protocol.

[33] The 1996 Protocol was different to the 1972 London Dumping Convention in recognising dumping also includes any abandonment or toppling at site of platforms or other man-made structures at sea, for the sole purpose of deliberate disposal. The point about including the seabed and subsoil was particularly important, as it finally answered a long term difference of opinion between the Parties during the 1980s (and whether nuclear waste placed into the seabed, as opposed to the ocean, was covered or not) over whether the London Dumping Convention did, or did not, cover this area. Annex V (LDC 8, 1984). Draft Resolutions on the Disposal into the Seabed of High Level Radioactive Wastes. Note article 10 of the Protocol.

[34] Articles 10(2), 10(3), 11(1) and 11(2). 1996 Protocol. For the agreed rules, see Annex VII. Compliance Procedures and Mechanisms for the London Protocol. (LDC 29/LP 2, 2007). Also, Annex VII. Amendments to the Base Text for Compliance Procedures. (LDC 28/LP 1, 2007). Reports from the Compliance Committee are included in every report of the Parties. The initial meetings were very much upon procedural considerations, such as reporting.

[35] Support for developing countries to help capacity building for marine pollution controls strategies in the marine environment can be traced to Recommendation 94 of the 1972 Stockholm Conference. This obligation carried over to

Somewhat uniquely for the period, the 1996 Protocol also adopted a precautionary approach that would restrict the allowance of contaminating substances into the marine environment 'where there was reason to suspect that harmful events may occur, even though stringent proof of a cause-effect relationship may be lacking'.[36] Accordingly, the Protocol agreed to pursue 'appropriate preventative measures ... when there is reason to believe that wastes or other matter introduced into the marine environment are likely to cause harm even when there is no conclusive evidence to prove a causal relation between inputs and their effects'.[37] Moreover, when making decisions about regulating waste into the ocean, rather than just base decisions on scientific or technical considerations (as was the basis in the 1972 Convention), the decision makers could also 'take into account legal, social and economic factors as appropriate'.[38]

The primary method through which the 1996 Protocol differed to the 1972 Convention was in reversing the dumping list.[39] What this meant was that, whilst in the past wastes would be listed which could not be dumped, now (unless an emergency)[40] the only wastes which could be

article 9 of the 1972 London Dumping Convention, which required support for low income countries, especially for human capacity building and necessary technical equipment for research and monitoring. The commitment was repeated in article 202 of the 1980 UNCLOS, and article 13 of the 1996 Protocol. Note also, Annex X. The Long Term Strategy for Technical Cooperation and Assistance under the London Convention and Protocol. (LDC 28/LP 1, 2007).

36 Report of the Thirteenth Consultative Meeting (1990, LDC 13/15) 18 Dec. Paragraph 3.29. See also Annex II (LDC 14, 1991). The Application of a Precautionary Approach in Environmental Protection Within the Framework of the London Dumping Convention.

37 Article 3(1). 1996 Protocol.

38 Article 22. 1996 Protocol. Note also, Annex III (LDC 15, 1992). Procedure for the Priority Consideration of Amendments to the Convention. The 1972 position is found in article 15(2) of the London Dumping Convention. Note also, Annex II (LDC V, 1980). Procedures for the Circulation of Proposed Amendments to the London Dumping Convention. For general commentary, see Sielen, A. (2009). 'The New International Rules on Ocean Dumping: Promise and Performance'. *Georgetown International Environmental Law Review*. 21: 295–340.

39 Annex III (LDC 15, 1992). Procedure for the Priority Consideration of Amendments to the Convention.

40 See article 8. For elaboration of this article, see Annex II. Procedures and Criteria for Determining and Addressing Emergency Situations under the London Protocol. (LDC 28/LP 1, 2007). For the build-up to this position, see Annex V. Draft Procedures and Criteria for Determining and Addressing Emergency Situations. (LDC 26, 2004); and again, see Annex V, in LDC 27, 2005.

dumped were those which were listed. This meant that the dumping of all waste, except that listed on Annex I to the Protocol, was prohibited.[41]

The wastes listed in Annex I which can be, prima facie, dumped are dredged material; sewage sludge; fish waste (or material resulting from industrial fish processing operations on the oceans or aquaculture, consisting of particles of flesh, skin, bones, entrails, shells or liquid stick-water); vessels and platforms or other man-made structures at sea; inert, inorganic geological material (waste resulting from mining activities); organic material of natural origin (usually, animal and vegetable matter predominantly of agricultural origin); and bulky items primarily comprising iron, steel, concrete and similarly unharmful materials for which the concern is physical impact, and limited to those circumstances where such wastes are generated. In 2007, this list was expanded to include carbon dioxide streams for disposal into sub-seabed geological formations.[42]

Further attempts to expand Annex I to include the geo-engineering practice of ocean fertilisation, whereby elements are added to the ocean in an attempt to stimulate biological activity which would produce beneficial results in terms of climate change, were the subject of strong debate between the Parties to the 1996 Protocol. Here, as at the time of writing, although such activities were still recognised as in certain exceptional instances acceptable if for legitimate scientific reasons, there were strong concerns over what would be considered permissible and how such dumping would be adequately managed.[43]

---

[41] Article 4. 1996 Protocol. For the development of this reverse listing process, see Annex III (LDC 17, 1994). Report of the Working Group on Reverse Listing.

[42] Annex VI/Resolution LP.1(1). On the Amendment to Include CO2 Sequestration in Sub-Seabed Geological Formations in Annex I to the London Protocol. (LDC 28/LP 1, 2007).

[43] See Annex III, Statement by Canada, and the three options on how to deal with Ocean Fertilisation (Annexes IV, V and VI), all in LDC 34/2012; Annex II. Statement by the Delegation of the United States on Ocean Fertilisation. (LDC 33/2011); Annex V/Resolution LC-LP.2. On the Assessment Framework for Scientific Research. (LDC 32/2010); Annex VI/Resolution LC-LP.1. On the Regulation of Ocean Fertilisation. (LDC 30/LP 2, 2008); Annex VI. Terms of Reference for the Scientific Group to Examine Ocean Fertilisation. (LDC 29/LP 2, 2007). For commentary, Kovac, C. (2013). 'Ocean Fertilisation: What Next?'. *Macquarie Journal of International and Comparative Environmental Law*. 9(2): 39–64; Chisholm, S. (2009). 'Ocean Fertilisation: Time to Move On'. *Nature*. 461(7262): 347; Rayfuse, R. (2008). 'Ocean Fertilisation and Climate Change: The Need to Regulate Emerging High Seas Uses'. *The International Journal of*

Even if waste is listed in Annex I and permissible to dump, assessments must still be undertaken.[44] The requirements for assessments were set down in Annex II by which the Parties must 'make further attempts to reduce the necessity for dumping'. These obligations include looking at the feasibility of using techniques that reduce or prevent ocean dumping. It was specifically recommended that a permit to dump wastes or other matter shall be refused if the permitting authority determines that appropriate opportunities exist to reuse, recycle or treat the waste – not in the ocean – without undue risks to human health, the environment or disproportionate costs. Following the initial waste audit, if ocean dumping is concluded to be the best choice, the chemical, physical and biological analysis of the waste and their impacts, in terms of their toxicity, persistence, accumulation and possible biotransformation in the ocean at the proposed dump site must be undertaken. Within these considerations, an Action List should be established for the highest level pollutants still found in the wastes of which it is permitted to dump (such as heavy metals within sewage sludge), and safety limits set. If the impacts and safety limits cannot be determined, the waste shall not be dumped. With wastes like old vessels, offshore installations or bulky items, it is assumed that all possible polluting material will be removed first, to the maximum extent.[45] If any of the Annex I wastes contain levels of radioactivity greater than *de minimis* (exempt) concentrations as defined by the International Atomic Energy Agency (IAEA) and adopted by the contracting Parties, they cannot be dumped.

The dump site must be carefully selected, taking into account the physical, chemical and biological characteristics of the water-column and the seabed; location of amenities; values and other uses of the sea; and economic and operational feasibility. Following this, an assessment of the potential effects (combining knowledge of the dump site, and the proposed waste) must be undertaken. From these considerations, ideal

---

*Marine and Coastal Law.* 23: 297–326; Freestone, D. (2008). 'Ocean Iron Fertilisation and International Law'. *Marine Ecology Progress Series.* 364: 227–33.

[44] This idea, in this context, originally appeared in 1992. See Annex IV (LDC 15, 1992). Waste Assessment Framework for the London Convention 1972.

[45] Ekins, P. (2006). 'Decommissioning of Offshore Oil and Gas Facilities'. *Journal of Environmental Management.* 79(4): 420–38; Hamzah, B. (2003). 'International Rules on Decommissioning of Offshore Installations'. *Marine Policy.* 27(4): 339–53; La Fayette, L. (1999). 'New Development in the Disposal of Offshore Instalations'. *The International Journal of Marine and Coastal Law.* 14(4): 523–49.

matches should be made. For example, with fish waste the target of where to dump should be where the likelihood of other species consuming the waste is strongest. Similarly, if engaged in trying to create benefits from the placement of artificial reefs from inert material, the locations should match.[46] Finally, scientific research (in general), monitoring of the particular sites, record keeping and compliance decisions must be undertaken of all sites where Annex I wastes are dumped into the ocean.[47]

These generic Annex II guidelines are then supplemented by (slightly) more specialised guidelines for the specific Annex I wastes for which permission to dump into the ocean is, prima facie, accepted. These guidelines cover inert, inorganic geological material;[48] organic material of natural origin;[49] bulky items (primarily iron, steel, concrete and similarly unharmful materials);[50] fish waste;[51] vessels;[52] man-made structures at sea;[53] carbon dioxide disposal in sub-seabed geological formations;[54] and sewage sludge and dredging.

---

[46] See UNEP (2007). *London Convention and Protocol/UNEP: Guidelines for the Placement of Artificial Reefs.* (UNEP, Nairobi).

[47] Annex II. 1996 Protocol. These were updated in 2008. See Annex III. Revised Guidelines for the Assessment of Wastes or Other Matter that May Be Considered Dumping. (LDC 30/LP 2, 2008). For the earlier work on these, see Annex II (LDC 19, 1997). Guidelines for the Assessment of Wastes or Other Matter that May Be Considered for Dumping; and Annex III (LDC 19, 1997). Format for the Waste Specific Guidelines in Relation to the Generic Guidelines.

[48] The guidelines were revised in 2008. See Annex IV. Revised Specific Guidelines for the Assessment of Inert, Inorganic Geological Material. (LDC 30/LP 2, 2008). For the earlier versions, see Annex VIII (LDC 24, 2001). Specific Guidelines for Assessment of Inert, Inorganic Geological Material. Note also, Annex VIII. Eligibility Criteria for Inert, Inorganic Geological Material. (LDC 28/LP 1, 2007).

[49] Annex IX (LDC 24, 2001). Specific Guidelines for Assessment of Organic Material of Natural Origin.

[50] The updated 2010 Guidelines for Assessment of Bulky Items (Annex II, LDC 32, 2010). For the earlier guidance on this, see Annex III (LDC 24, 2001).

[51] The updated 2012 Guidelines for the Development of Action Lists and Action Levels for Fish Waste (Annex II, LDC 34) complement Annex V (LDC 24, 2001). Specific Guidelines for Assessment of Fish Waste.

[52] Annex VI (LDC 24, 2001). Specific Guidelines for Assessment of Vessels.

[53] Annex VII (LDC 24, 2001). Specific Guidelines for Assessment of Platforms or Other Man Made Structures at Sea.

[54] Annex VIII. Specific Guidelines for the Assessment of Carbon Dioxide for Disposal into Sub Seabed Geological Formations (LC 34/2012). Also, on the same topic, the earlier Annex IV (LDC 29/LP 2, 2007).

## 5. THE CHANGE IN PRACTICE

Four examples can be produced of the way that dumping waste into the oceans has changed in practice between 1972 and 1996. The first two examples, of nuclear and industrial wastes, show a movement from regulation to prohibition. The second two, of dredging and sewage sludge, show a movement from minimal oversight to much greater regulation.

Originally, the dumping of industrial waste, unless it triggered the considerations found in Annex I of the 1972 Convention, was permissible, and was commonly seen as an Annex III/White List activity – with minimal restrictions. After 1996, following a groundswell of opinion against the dumping of industrial waste into the ocean in the early 1990s, the Parties to the London Dumping Convention agreed to phase this practice out by 1995.[55] Although there were teething difficulties with this prohibition, it was followed by the 1996 Protocol, in which the dumping of industrial waste into the ocean was clearly omitted as a permissible type of activity.[56]

A similar type of progress can be seen with nuclear waste. This waste stream was mentioned in both Annex I and II of the original 1972 London Dumping Convention. With regard to prohibited dumping, Annex I of the London Dumping Convention covered 'high level radio-active wastes or other high level radio-active matter'.[57] Annex II allowed regulated dumping of certain types of waste that were 'radio-active wastes or other radio-active matter not included in Annex I'.[58] That is, low-level and medium-level nuclear types of waste were, prima facie, acceptable for regulated dumping into the ocean, but high-level wastes were not. In both the Annex I and II categories, deference was given to the IAEA, who continually gave guidance on what would be, given that radionuclides were already omnipresent (both natural and anthropogenic)

55 Annex IX (LDC 13, 1990). Phasing Out Sea Disposal of Industrial Waste. Also, Annex III (LDC 16, 1993). Amendments to the Annexes of the LDC Concerning Phasing Out Sea Disposal of Industrial Waste.

56 Report of the 21st Consultative Meeting. LC21/13 (1999). Paragraph 5.12; Report of the 24th Meeting of the Scientific Group of the London Dumping Convention. LC/SG 24/11. 24 July 2001. Paragraph 4.30.

57 Paragraph 6 of Annex I.

58 Paragraph c of Annex II.

in the ocean, and a minimum (*de minimis*) level of radioactivity, which would thus be exempt from the Convention.[59]

As scientists argued over the *de minimis* basis, a number of countries came to argue against the dumping of any nuclear waste into the ocean (high, medium or low-level) or seabed. Opposing this coalition, a minority of other countries argued that carefully monitored dumping of low-level nuclear waste into the ocean was appropriate. These matters came to a head in 1983 when a majority of countries forced through an amendment to the Annexes, with a de facto moratorium until it could be proven that any dumping of radioactive wastes and other radioactive matter at sea … 'will not harm human life and/or cause significant damage to the marine environment'.[60] This approach, seeking to require proof of harmlessness as a precondition for dumping, was revolutionary.

Before this matter could be fully resolved, the Cold War began to thaw, and a new generation of environmental thinking came into being, accepting that the London Dumping Convention also covered both the ocean and the seabed, and that the prohibition of dumping low-level nuclear waste was covered in both areas. This new consensus was recorded at the 1992 Earth Summit,[61] and was followed through by the domestic actions of key countries like the United States. Countries that had technical difficulties trying to manage this goal, such as Russia, were actively assisted by being helped to develop the necessary facilities to successfully deal with their nuclear waste, without having to dump it into the ocean. With such assistance, Russia could come to accept Annex I of the 1996 Protocol which stipulated 'materials … containing levels of

---

59 Annex VI (LDC 21, 1999). Guidelines for the Application of the De Minimis Concept under the London Convention, 1972; Annex X (LDC IV, 1979). Revised IAEA Definition Concerning Radioactive Wastes. Also, Templeton, D. (1997). 'Research Needed Relative to Radiological Assessment of the Dumping of Radioactive Wastes in the Oceans'. *Marine Pollution Bulletin*. 35(7): 374–80; Calmet, D. (1991). 'Radioactive Waste and Ocean Dumping: The Role of the IAEA'. *Marine Policy* 15(6): 413–20.

60 In sequence, the important decisions were: Annex IV (LDC 8, 1984). Terms of Reference for the Review of Scientific and Technical Considerations Related to Dumping of Radioactive Waste at Sea; Annex V (LDC 8, 1984). Resolutions on the Disposal into the Seabed of High Level Radioactive Wastes; Annex IV (LDC 9, 1985). Dumping of Radioactive Wastes at Sea; Annex XI (LDC 10, 1986). Studies and Assessments on the Disposal of Nuclear Waste in the Ocean; and Annex XII (LDC 10, 1986). Statements Expressing Reservations to Annex XI.

61 See paragraph 5 of chapter 22 of Agenda 21 (1992). UNCED. Doc A/CONF.151/4.

radioactivity greater than *de minimis* (exempt) concentrations as defined by the IAEA … shall not be considered eligible for dumping'.[62]

Two examples of the change of practice, from little oversight to regulation between the 1972 Convention and the 1996 Protocol, can be seen with dredging and sewage sludge. Originally, the view of the international community was that the dumping of dredging waste was considered an Annex III/White List activity under the 1972 London Dumping Convention and largely unregulated. This unregulated approach changed in 1996 with the Protocol, under which it became regulated, and assisted with the development of guidelines around this area.[63] Specifically, the Revised Guidelines for the Sampling of Sediment Intended for Disposal at Sea were adopted in 2004 as was the Guidance for Action Lists and Action Levels for Dredged Material, with the latter setting down recommendations for screening candidate wastes and establishing upper and lower levels of concern for the projected dump site. The end result is a regime in which for dredged material, the goal is to identify and attempt to control the original sources of contamination (as in, where the hazards deposited in the dredgings derived from, and to try to solve that problem as a first step). Thereafter, an evaluation of the need for dredging and disposal, the identification of any possible beneficial use of the material, as well as any possible negative consequences, are to be critically assessed against the specific material that has been dredged (and exactly how polluted it is, or is not) and the proposed sites that it may be dumped at, and the costs and benefits of these specific sites.

---

[62] For the debate in the LDC, see Annex V (LDC 12, 1989). Resolution on the Disposal of Radioactive Wastes into Sub-Sea-Bed Repositories Accessed from the Sea; Annex VII (LDC 13, 1990). Disposal of Radioactive Wastes into Sub-Sea-Bed Repositories Accessed from the Sea; Annex V (LDC 16, 1993). Amendments to the Annexes Concerning Radioactive Wastes; Report of the 27th Consultative Meeting of the Parties to the London Dumping Convention. LC/16. 16 Dec. 2005. Paragraph 11; Annex VI (LDC 17, 1994). Status Report on the Treatment and Disposal of Low-Level Radioactive Wastes Derived from Maritime Activities. See also Annex III, The White House Statement, attached to Annex VI (in the same Report). For commentary on this, see Vandenbosch, R. (2007). *Nuclear Waste Stalemate*. (Utah UP, Salt Lake). 13; Alley, W. (2011). *Too Hot to Touch: The Problem of High Level Nuclear Waste*. (Cambridge UP, Cambridge). 39, 44–5; Ringius, L. (1997). 'Environmental NGOs and Regime Change: The Case of Ocean Dumping of Radioactive Waste'. *European Journal of International Relations*. 3: 61–96; Anon. (2004). 'Nuclear Subs Plea'. *New Scientist*. 22 May. 5.

[63] Annex II (LDC 10, 1986). Guidelines for the Application of the Annexes to the Disposal of Dredged Material.

Quite simply, it is no longer assumed that it is fine to dump any dredged material in any location desired.[64]

It has long been the practice to dispose of sewage (ideally treated to some basic level) into the ocean.[65] Such solutions were seen as purely domestic concerns until 1972, when the London Dumping Convention recognised the disposal of sewage sludge into the oceans as a concern, albeit, a minimal one. At this point, it was considered an activity on the White List, with minimal regulation, assuming that the sludge had some treatment and was dumped in suitable locations. In fact, it was stated categorically that Annex I – for waste streams that were absolutely prohibited – did not apply to sewage sludge, even if it contained some of the other absolutely prohibited wastes for dumping. Rather, such sewage wastes had to be 'managed' within the goals of regulated dumping. Specifically, if the depth, temperature, salinity, presence of oxygen, currents, seasonal flows and storm patterns were favourable, and there were no particularly susceptible species in the way, it was assumed that the dumping of such sludge was acceptable.[66] The acceptance of such practices only began to change when greater scientific understanding of the elevated concentrations of heavy metals, human enteric viruses, chloro-biphenyls and polycyclic aromatic hydrocarbons as found in ocean dump sites, long after the sludge had been dumped, caused many to question the safety of this waste dumping.[67]

---

[64] Annex VII. Guidelines for the Sampling and Analysis of Dredged Material. (LDC 26, 2004). Also, International Maritime Organisation (2002). *Guidance for the Development of Action Lists and Action Levels for Dredged Material.* (IMO, London). Annex III (LDC 24, 2001); Specific Guidelines for Assessment of Dredged Material. Annex II (LDC 24, 2001); and Development of Generic Guidelines for the Selection of Physical, Chemical and Biological Variables for the Assessment of Dredged Material.

[65] Benidickson, J. (2007). *The Culture of Flushing: A Social and Legal History of Sewage*. (UBC Press, Vancouver). 110, 118–19; Colten, C. (1996). *The Road to Love Canal: Managing Industrial Waste Before the EPA*. (Austin, Texas UP). 76; Wastkerm T. (1981). 'Ocean Dumping under the London Dumping Convention in the United States'. *Chemosphere.* 10(6): 659–81.

[66] Smith, J. (1987). *Marine Treatment of Sewage and Sludge*. (Telford, London); Dart, M. (ed.) (1981). *Disposal of Sludge to Sea.* (Pergamon, London); Wastler, T. (1981). 'Ocean Dumping Permit Program under the London Dumping Convention'. *Chemosphere.* 10(6): 659–60.

[67] Eljarrat, E. (2001). 'Evaluation of Dioxin Contamination in Sewage Sludge Discharges on Coastal Sediments'. *Water Research.* 35(11): 2799–803; Comber, S. (1995). 'Behaviour of Trace Metals in Sewage Sludge after Dispersal in Seawater'. *Environmental Technology*. 16(5): 230–43; Takizawa, M. (1993). 'Near-Bottom Pelagic Bacteria at a Deep-Water Sewage Sludge Disposal Site'.

Due to such concerns, some developed countries came to completely prohibit the dumping of sewage sludge into the ocean. These domestic trends were reflected at the international level, where a majority of the Parties set the aim in 1992 'to ultimately alleviate the need for sewage sludge disposal at sea'.[68] However, this goal of total prohibition of the dumping of treated sewage sludge into the ocean was not accepted by the international community in either Agenda 21[69] or the 1996 Protocol, where the goal was to improve the regulation, not prohibition of, the dumping of sludge into the ocean. Even area specific regimes which dealt with the most sensitive of all oceans, such as around Antarctica, recognised this regulated, as opposed to prohibited, approach.[70] The 1996 Protocol continued this trend, recognising the dumping of sludge as a permissible activity, if done in accordance with clear procedures.[71] These procedures are a combination of Annex II of the 1996 Protocol and the Specific Guidelines for the Assessment of Sewage Sludge.[72] There are three steps to this procedure.

---

*Applied Environmental Microbiology.* 59(10): 3406–10; Goyal, S. (1984). 'Human Pathogenic Viruses at Sewage Sludge Disposal Sites'. *Applied Environmental Microbiology.* 48(4): 758–63; Bothner, M. (1994). 'Sewage Contamination in Sediments Beneath a Deep-Ocean Dump Site'. *Marine Environmental Research.* 38(1): 43–59; Costello, M. (1994). 'Toxicity of Sewage Sludge to Marine Organisms: A Review'. *Marine Environmental Research.* 37(1): 23–46; Rodger, G. (1992). 'The Recovery of a Sewage Sludge Dumping Ground'. *Science of the Total Environment.* 119(1): 1–11; Angelidis, A. (1991). 'The Segregation of Metals During the Ocean Dumping of Sludge'. *The Science of the Total Environment.* 102: 129–45; Gibbs, R. (1989). 'Effect of Sludge Digestion on Metal Segregation During Ocean Dumping'. *Marine Pollution Bulletin.* 20(10): 503–10; Chapman, D. (1986). 'The Distribution of Metals in Sewage Sludge and Their Fate after Dumping at Sea'. *Science of the Total Environment.* 48(1): 1–11; Webster, L. (1978). 'Organic Contaminants in the Firth of Clyde following the Cessation of Sewage Sludge Dumping'. *Water.* 12(8): 535–45.

68 Annex III (LDC 15, 1992). Procedure for the Priority Consideration of Amendments to the Convention; and Annex VI (LDC 14, 1991). Concerning the Protection of the Oceans and All Kinds of Seas.

69 Agenda 21 (1992). UNCED. Doc A/CONF.151/4. Chapter 27.

70 That is, after initial dilution and rapid dispersal occurred for stations below 30 people, and treatment before ocean dumping, if over 30 people, the standard practice is the sludge is repatriated. See in particular, Annex III of the Madrid Protocol, on Waste Disposal and Waste Management. Also, Connor, M. (2008). 'Dealing with Sewage in Antarctica'. *Polar Record.* 44(2): 165–71.

71 Article 4 and Annex I. 1996 Protocol.

72 Annex IV (LDC 24, 2001). Specific Guidelines for Assessment of Sewage Sludge.

First, a waste prevention audit must be undertaken, into the type, amount and relative hazard of the sewage sludge. The focus here, rather than trying to stop the overall amount, is to identify and control any supplemental pollutants in the sludge. Second, the alternatives to dumping of sewage sludge at sea, such as putting it back onto the land, incineration or disposal into suitably designed landfills, should be considered. Fundamentally, it is recommended that ocean dumping of sludge should not occur if the alternatives can be done safely and without disproportionate economic cost. Finally, an assessment of the potential effects must be undertaken. This assessment is composed of two parts – the waste, and the dump site. With regard to the sludge, a detailed analysis of the chemical, physical and biological properties of the waste must be undertaken to see exactly how hazardous it is. Priority is meant to be given to identifying toxic, persistent and bio-accumulative substances (especially heavy metals and organohalons) within the sludge. With regard to the site, each proposed dump location must be evaluated in light of its biological (such as depth, temperature, turbulence of the water, potential links with oxygen demand, nutrient fluxes and potential eutrophication and vulnerable species – especially those consumed by humans), social (such as closeness to shorelines, bathing beaches, areas of cultural significance, and so on) and technical (size, capacity, contaminant mobility) considerations. Thereafter, if approval is given, close monitoring of the waste, operation and site must be undertaken.

## 6. THE LIMITS OF CHANGE

Whilst the progress made with the 1996 Protocol and the forthright restrictions on the traditional practices of ocean dumping are to be welcomed, their overall effectiveness in protecting the health of the ocean have been limited by other developments elsewhere, and waste streams that the Parties to the London Dumping Convention and the 1996 Protocol have much greater difficulty controlling. The foremost example of this is the plastic in the ocean. As discussed in Chapter 2, this is a problem of such magnitude that the international community has committed itself to 'take action by 2025, based on collected scientific data, achieve significant reductions in marine debris to prevent harm to the coastal and marine environment'.[73]

[73] Rio+20. A/CONF.216/XX. Paragraph 163.

The response of the international community, to date, has been through dividing the source of plastic pollution in the ocean into two parts. The first source of plastic is from direct dumping into the seas. It has been agreed since 1972 that the dumping of plastic waste into the ocean was an Annex I – absolutely prohibited – activity. Of this prohibition, there has never been any disagreement by the Parties to the London Dumping Convention.[74] The second source of plastic in the ocean originates from ocean going vessels. The control of this waste coming from ships (as opposed to a dedicated waste stream) is dealt with by the 1973/78 International Convention for the Prevention of Pollution from Ships (MARPOL).[75] Annex V of this Convention specifies the distances from land and the manner in which the garbage may be disposed of. The requirements are much stricter in a number of the recognised 'special areas'. In addition, the most important feature of the Annex is the complete ban imposed on the dumping into the sea of all forms of plastic, irrespective of distance or location. Similar arrangements exist with some regional fisheries organisations, and the obligations they impose to control synthetic/plastic fishing products such as in the Code of Conduct for Responsible Fisheries of the Food and Agricultural Organization of the United Nations (FAO).[76] From such a concern, measures to address lost or abandoned gear and related marine debris have been adopted by, inter alia, the European Community, New Zealand, Norway, Saudi Arabia and the United States. The United Nations General Assembly has complemented this area, calling on all states to integrate the issue of marine debris into national strategies dealing with waste management in the coastal zone, ports and maritime industries.[77] In addition, a number of specific fisheries-related conventions, reflecting the concerns of the FAO's Committee on Fisheries (COFI),[78] such as the 1980 Convention on the Conservation of Antarctic Marine Living Resources, have banned certain fishing materials which would otherwise be disposed of in the

---

74 See paragraph 4 of Annex I of the 1972 London Dumping Convention. For affirmation of this position, see Annex VI (LDC 9, 1985). Environmental Hazards Caused by the Disposal at Sea of Persistent Plastics.

75 Reprinted in Kiss, A. (ed.) (1980). *Selected Multilateral Treaties in the Field of the Environment.* (UNEP, Nairobi). 320.

76 FAO (1995). *Code of Conduct for Responsible Fisheries.* (FAO, Rome). Section 8.4.6. Also, FAO (1994). 'Expert Consultation on the Code of Conduct for Responsible Fishing'. FIIT/R506. Paragraphs 17, 23.

77 General Assembly Resolution (2010). 64/71. Oceans and the Law of the Sea. Paragraph 121.

78 COFI (2007). Report of the 27th Session of the Committee of Fisheries (COFI/FAO). xiii.

ocean. They have also required the marking of all fishing gear with the letters of the boat they belong to, so that any plastic litter can be traced to its last point of origin.[79]

The problem here is that the plastic waste in the ocean is continuing to grow despite all the restrictions by the London Dumping Convention, the MARPOL and associated fisheries regimes. This is because the vast majority of the plastic coming into the ocean is not coming from these sources. Rather, it is coming from land sources, especially as litter or from uncontrolled processes, such as washing machines, which may be a dominant source for the creation of micro-plastic (which then makes its way into the waterways), from where it has to be controlled by other means. The response of the international community in this area has been through initiatives such as the Global Programme of Action for the Protection of the Marine Environment from Land-Based Activities, and other local and regional initiatives which are seeking to address this issue. All of these soft initiatives have called for reductions of the problem of plastic debris at source and/or effective waste management strategies, such as recycling, which capture this waste before it gets into the ocean. The problem, as the earlier chapters in this book hope to have shown, is that this response is very limited and is highly unlikely to address the source of the difficulty of the rapid accumulation of plastic in the ocean.[80]

---

79 CCAMLR (1998). 'Marking of Fishing Vessels and Fishing Gear'. Conservation Measure 10–01; CCAMLR (2008). 'General Environmental Protection During Fishing'. CCAMLR, Conservation Measure 26–01. See also, UNEP/FAO (2008). 'Abandoned, Lost or Otherwise Discarded Fishing Gear'. *UNEP Regional Seas Report*. 185: xiv–xviii.

80 Vo, S. (2013). 'Lessons Learned in Coastal Habitat and Land Based Pollution Management'. *Ocean and Coastal Management*. 85: 230–43; UNEP (2012). *Global Environment Outlook 5*. (Progress Publishers, Malta). 173; Moore, C. (2012). *Plastic Ocean*. (Avery, NYC). 68; Jones, T. (2011). 'Home Washing Machines: Source of Harmful Ocean Microplastic'. *Ecology, Environment and Conservation*. 11: 193–204; Ariza, E. (2009). 'Seasonal Evolution of Beach Waste and Litter'. *Waste Management*. 28: 2604–13; Bravo, A. (2009). 'Anthropogenic Debris on Beaches in the SE Pacific'. *Marine Pollution Bulletin*. 58: 1718–26; Vander, Z. (2008). 'The Protection of the Marine Environment from Land-Based Pollution and Activities'. *The International Journal of Marine and Coastal Law*. 23(3): 423–40; Li, D. (2004). 'Ocean Pollution from Land Based Sources'. *Ambio*. 33(1): 107–20.

## 7. CONCLUSION

When it was impractical to incinerate or dump waste into the land, the ocean was always the preferred target for waste disposal. Whilst domestic restraints in this area began in the developed world in the early part of the 20th century, international ones did not emerge until the 1970s. The international restraints, as found through the London Dumping Convention, were the subject of intense debate for over a decade, during which time the dumping of most types of waste into the ocean was the practice, not the exception. Only the most hazardous of wastes were kept out. Thus, whilst high-level nuclear wastes could not be dumped into the ocean, low-level wastes could be.

This approach was completely turned around in 1996 when the Protocol to the London Dumping Convention reversed the assumption that it was safe to dump anything into the ocean, and only a few specific waste streams could be deposited into the seas. Accordingly, many of the practices that were previously acceptable, became unacceptable. This was particularly obvious with regard to the dumping of either industrial or nuclear wastes. However, despite the progress being made in this area by regimes like the London Dumping Convention, the greatest of all waste problems emerging in the oceans – plastic – cannot be solved by such regimes, as the source of the problem is on land. Accordingly, it is only if methods such as technological change, producer responsibility, recycling and much more effective waste management are used to tackle waste plastic will the problem of this waste stream entering the oceans be solved.

# 8. Human waste

## 1. INTRODUCTION

This chapter covers the topic of the waste stream that comes directly from the human body. It has been put into a separate chapter, as this particular type of waste causes more deaths than all of the other waste streams put together. Despite these impacts, and the concerted efforts of scientists and policy makers to effectively deal with this problem for over 150 years, the progress has become very divided between the developed and the developing world, and where each region is with regard to dealing with human waste as a first, second or third generation type of problem.

The first generation problem was about getting the waste out of human communities. The second generation has been about treating it, and the third, final disposal. Despite strong international interest, the success in achieving the primary goal is mixed, with strong divisions being evident between developed and developing countries. The same can be said for dealing with the international control of disposing treated human waste into freshwater systems. This is not the case for the disposal of treated human waste into the ocean, where the international regulation is much more robust, and as seen in the previous chapter, a trend is developing whereby this practice is becoming less popular. Conversely, the options that are becoming popular for final disposal are landfill, incineration or recycling. Of all these three, each has both limits and benefits, although the last option, involving the recycling of biosolids, appears to be the best use of the waste stream, as a potential resource.

## 2. BEGINNINGS

The first type of waste that humanity came into contact with was that which is emitted from each of our bodies. Of these two streams, urine is virtually free of hazardous material. The same is not true of excreta. This contains a large number of pathogens which are dangerous to health. Millions of viruses and bacteria are present in a typical gram of faeces.

Many of these are simply by-products of the human system and are not dangerous. Others, including many of the same bacteria in a pathogenic form, are the cause of many mild and serious types of fever and stomach problems which beset millions of people. Diarrhea is the most important excreta-related problem, but intestinal worms, whose eggs mature in faeces and enter the body through feet or the mouth, can also do great damage.[1]

From the pathogens related to human excrement, an estimated 1.8 million deaths occur every year. This figure includes cholera, which by the spread of toxic bacteria through water kills about 90,000 people per year. Diarrheal diseases take a further 1.6 million human lives. Ninety percent of these deaths are of children under five. Most of these deaths occur in developing countries, where over half a billion children have no toilet facilities – out of a global total of 2.6 billion people on the planet who lack adequate sanitation. Eighty-eight percent of diarrheal disease is attributed to unsafe water supply and inadequate sanitation and hygiene. Additional pathogens associated with untreated, or poorly treated, wastewater include *Salmonella*, *Campylobacter*, *Escherichia coli*, *Giardia*, *Cryptosporidium* and viruses. These may also cause problems to surrounding populations. This is especially so with gastrointestinal problems, but also respiratory, ear and eye problems are associated with pathogens released into water bodies. Evidence suggests that improved sanitation can reduce diarrhea morbidity by 45 percent, and the simple act of washing hands at critical times can reduce the number of diarrheal cases by an additional 35 percent.[2]

The ancient Babylonians were the first to begin experimenting with technologies designed to manage human waste. Four thousand years ago some of these people had dwellings which had toilets in them. These early toilets were almost identical to those in much of the Middle East in contemporary times, with a small oblong hole in the floor (without a

---

1 Black, M. (2008). *The Last Taboo: Opening the Door on the Global Sanitation Crisis*. (Earthscan, London). 78–83.

2 Editor. (2014). 'Water and Sanitation: Addressing Inequalities'. *The Lancet*. 38(9926): 1359; Tarrass, F. (2012). 'The Effects of Water Shortages on Health and Human Development'. *Perspectives in Public Health*. 132(5): 240; Ali, M. (2012). 'The Global Burden of Cholera'. *Bulletin of the World Health Organization*. 90: 209–18; Giusti, L. (2009). 'A Review of Waste Management Practices and Their Impact on Human Health'. *Waste Management*. 29: 2227–39; Buttenheim, A. (2008). 'The Sanitation Environment in Urban Slums: Implications for Health'. *Population and Environment*. 30: 26–47; Bartram, J. (2005). Focusing on Improved Water and Sanitation. *The Lancet*. 365: 810–12.

seat) and some kind of flushing of water for the offensive material. The Old Testament added rules for the disposal of sewage, in terms of going from the living area and burying the waste, and a little later, around 1000 BCE, two independent systems of street drainage were operating in Jerusalem. One for sewage and one for wastewater.[3]

The first sewers in Rome were built between 800 and 735 BCE, and therefore antedate the first aqueducts that were constructed between 440 and 487 BCE. This sewage system was apparently so complete that every street in the city was drained by a branch to the Tiber. Similar models were followed in many other Roman settlements, with open spaces for the waste to congregate being the end point, if waterways to exit the sewage did not exist. These processes were supplemented by theological support and technological knowhow.[4] The Roman disposal systems were supplemented by a flow of water, drains and raised sidewalks. The Romans also utilised cesspits, which were commonly placed inside houses. Once filled, the cesspits were emptied by specialist manure merchants, who would go on to sell the waste as fertiliser. Cesspits for ground level dwellings were supplemented with sewers for some public collections of human excreta. The difficulty with the Roman system was that sewage removal mechanisms only existed for ground level dwellings. Anything above the ground was often jettisoned from windows from above. To provide against accidents due to this cause, the Dejecti Effusive Act was passed, which gave damages against a person who threw or poured anything from a height onto a public road or path. Damages were only recoverable for physical damage to the body incurred during daylight, not the inconvenience or nuisance of being covered in human waste if the incident occurred at night. To this day, archeologists continue to recover ancient signs warning people of where to stand, sit or traverse at their peril.[5]

With the collapse of the western portion of the Roman Empire, so too the knowledge and technical ability of how to manage sewage in urban environments disappeared. This only began to reappear in the 13th

---

[3] Hodge, T. (2012). *Roman Aqueducts and Water Supply.* (Bristol Classics, London). 334–5; Rathje, W. (1992). *Rubbish: The Archaeology of Garbage.* (HarperCollins, NYC). 19.

[4] The Romans had three gods in this area, dealing with the respective problems of ordure – Stercutius; toilets – Crepitus; and sewers – Cloacina.

[5] Hodge, T. (2012). *Roman Aqueducts and Water Supply.* (Bristol Classics, London). 334, 336, 337, 344–5; Strasser, S. (1999). *Waste and Want: A Social History of Trash.* (Holt, NYC). 15; Cosgrove, J. (1909). *A History of Sanitation.* (Standard, Pittsburgh). 80–82, 85.

century, when many European cities, possibly enlightened by the practices brought back from the Middle East and Byzantium, began to manage human waste via the provision of sewers. In fact, it is from this period that the word 'sewage' derives, being from the word 'sewer', which finds its roots in the Anglo-Norman *sewere* (watercourse) and the Old French *sewiere* (overflow channel for a fishpond).

Florence and Venice set up sanitary commissions in the 1350s. In England, Henry VIII was aware of the 'great damages and losses' attributed to the 'outrageous surges' of sewage waste going into the rivers, but also 'spilling onto private and common lands'. Accordingly, Henry created a Commission of Sewers which was authorised to enforce repairs on existing sewers, and made expedient decisions to help the flow of excrement. However, progress was slow and it was not until 1700 that large cities like London had a steady water supply. This was not unusual. Comparable cities of the day, such as Paris or Boston, did not begin the construction of anything even beginning to resemble a comprehensive sewer network until the end of the 17th century. Even with these few systems that were emerging, the laws that existed worked in what are now considered counter-intuitive ways. That is, in London, the discharge of any water except kitchen slops into the drains was prohibited until 1815.[6]

It was only moving into the 19th century that policy makers began to prioritise the need to better manage human waste in the urban environment. Although they were long aware of the problems of filthy environmental conditions caused by the lack of adequate water supplies, drainage and sewers, there had not been any major epidemics that impacted upon large sections of the population. This all changed in the first decades of the 19th century. Populations in the United States suffered outbreaks of yellow fever and cholera in 1832, 1849, 1866, 1873 and 1878 – with the last one killing over 10,000 citizens. However, it was the first two outbreaks, which overlapped with those in Britain, that were at the core of a fundamental shift in the way that human waste was dealt with, after London faced epidemics of cholera in 1832, 1848, 1849 and

---

[6] The law in question was the Act Concerning a Commission of Sewers. 3 Statutes of the Realm 368. 1509–1545. For the law on pollution, see the Act of 1231 of Frederic II. Reprinted in Herlihy, D. (ed.). *The History of Feudalism.* (Harper, NYC). 270–71. For commentary, see Mumford, L. (1946). *The Culture of Cities.* (Secker, London). 42–6; Strasser, S. (1999). *Waste and Want: A Social History of Trash.* (Holt, NYC). 50–59; Cosgrove, J. (1909). *A History of Sanitation.* (Standard, Pittsburgh). 85, 88–93, 100–107; Abert, J. (2013). *An Environmental History of the Middle Ages.* (Routledge, London). 62–7.

1853. These waves of disease killed tens of thousands of citizens. The 1849 outbreak alone claimed the lives of over 14,000 people.[7]

In the 1840s the city of London was home to some 2.5 million citizens. The vast majority of these cooked, washed, slept and defecated – often with the assistance of the latest must have device – the modern toilet – as finalised by the celebrated Thomas Crapper, in their private dwellings. The problem was that the new toilets, like the waste management systems from hundreds of years before them, led directly to the cesspits, not to any sewage system. London by this point had over 200,000 cesspits. Thousands of these overlapped with local water supplies. When people could no longer afford the high prices demanded by the 'night-soil' men to empty the cesspits, the cesspits overflowed. When cholera entered into the population, which was passed into excrement that then leached its way into the water supply, the epidemics exploded as sections of the population began drinking a concoction of cholera carrying sewage. At this point, a number of scientists and doctors, and John Snow in particular, managed to understand the negative mix of population density, location of the disease, heat and freshwater supplies. The solution to the problem, quite simply, was to keep the disease carrying sewage separate from fresh water, or as the Poor Law Commission of 1842 noted 'the introduction of cheaper and more efficient modes of removing all noxious refuse from the towns, are operations for which aid must be sought from the science of the civil engineer'.[8] This goal was orchestrated under the 1848 Public Health Act. This legislation was created:

> [F]or improving the sanitary condition of towns and places in England and Wales … it is expedient that the supply of water to such towns and places, and the sewerage, drainage, cleansing, and paving thereof, should … be placed under one and the same local management and control.[9]

---

[7] Black, M. (2008). *The Last Taboo: Opening the Door on the Global Sanitation Crisis*. (Earthscan, London). 14–15; Hoy, S. (1996). *Chasing Dirt: The American Pursuit of Cleanliness*. (OUP, Oxford). 26–9.

[8] Poor Law Commissioners (1842). *An Enquiry into the Sanitary Condition of the Labouring Population of Great Britain*. (Clowes, London). 7. Also, Tynan, N. (2013). 'Nineteenth Century London Water Supply: Processes of Innovation and Improvement'. *The Review of Austrian Economics*. 26(1): 320–40; Cadbury, D. (2003). *Seven Wonders of the Industrial World*. (4th Estate, London). 160–65.

[9] The 1848 Public Health Act. 11 and 12 Vict. c. 63. As reprinted in Lumley, W. (1859). *The New Sanitary Laws*. (Shaw Printers, London). 15.

This Act provided a raft of previously unseen measures to deal with sewage. At the core of these was the creation of a new board that was given the power to purchase, create, maintain and enforce standards for the management of sewage. These standards ranged from requiring all new houses and places of employment to have sufficient water closets, through to all such new and existing places being adequately connected to the necessary sewers. In practice, this meant over 1,000 miles of coordinated and comprehensive sewers, laced with 82 miles of intercepting drains, which would stretch across 80 square miles of London. This same pattern was then replicated during the second half of the 19th century in other parts of Europe, the United States, Japan and associated colonies. By the end of the 19th century most of the emerging industrialised countries were making great strides in getting a good supply of clean water into their cities and the sewage out.[10]

The difficulty that faced these communities a few decades later in the beginning of the 20th century was that although they had recognised and dealt with the problem of getting human waste out of human populations and away from freshwater sources, they did not really know what to do with the human waste next. Indeed, by the turn of the 20th century, the great sewers of London were now flushing thousands of gallons of collected – but untreated – London sewage per day, into the Thames. This volume of sewage (in addition to other industrial pollutants), constituting roughly one-sixth of the flow of the water, nearly killed the river in its entirety. The same pattern was replicated in parts of Europe and the United States. Alternatively, the waste was taken to the ocean and either dumped in a raw, or partly dried, form.[11]

With such a situation, it became evident that more advanced forms of management had to be developed to mitigate the problems. In the case of sewage, the technical answers arose from a series of British Royal Commissions that were tasked to investigate the best ways to make the

---

[10] Fee, E. (2005). 'The Public Health Act of 1848'. *Bulletin of the World Health Organization.* 83(11): 866; Cadbury, D. (2003). *Seven Wonders of the Industrial World.* (4th Estate, London). 170–78; Halliday S. (1999). *The Great Stink of London: Sir Joseph Bazalgette and the Cleansing of the Victorian Metropolis.* (Sutton, Gloucestershire). 130; Hoy, S. (1996). *Chasing Dirt: The American Pursuit of Cleanliness.* (OUP, Oxford). 14, 61–5; Cosgrove, J. (1909). *A History of Sanitation.* (Standard, Pittsburgh). 88–93, 100–107.

[11] Benidickson, J. (2007). *The Culture of Flushing: A Social and Legal History of Sewage*. (UBC Press, Vancouver). 110, 118–19; Colten, C. (1996). *The Road to Love Canal: Managing Industrial Waste Before the EPA*. (Austin, Texas UP). 76; Wastkerm, T. (1981). 'Ocean Dumping under the London Dumping Convention in the United States'. *Chemosphere.* 10(6): 659–81.

waste, which was being quickly taken out of the cities, safe. The answer was to treat and carefully dispose of it. The conclusions of these Commissions set the standards for the treatment of sewage which continued throughout the 20th century, and into in the 21st.

Specifically, by 1915 it had been concluded that sewage had to have, at least, a primary and a secondary treatment. The primary treatment involved the removal of settable organic solids by sedimentation and the removal of materials that float by skimming. Secondary treatment followed. This sought to remove the dissolved and non-settling organic solids from the primary effluent by using microbial populations. By the 21st century, two further treatments to remove grit and pathogens would be added. In contemporary settings, these treatments may be aerobic, anaerobic or facultative. The end result of these three or four treatments of sewage is that water can be released back into the natural environment, whilst the end product of a type of sludge (a semi-solid material) needs to be carefully disposed of. The reason why care is required with the dealing of sludge is because this organic-rich waste was brought in with other aqueous waste streams which have captured urban, domestic and/or industrial pollutants that have all gone down the sewer or through the human body to begin with. The end result is that sludges often contain pollutants that have accumulated during wastewater treatment. These include dozens of metals (including many heavy metals in particular), pharmaceuticals, steroids, hormones, pathogens, flame retardants and organic pollutants, including polychlorinated dibenzo-p-dioxins (PCDDs), dibenzofurans (PCDFs) and polybrominated diphenylethers (PBDEs).[12]

---

[12] Mailler, R. (2014). 'Priority and Emerging Pollutants in Sewage Sludge'. *Waste Management*. 34(7): 1217–35; Cincinelli, A. (2012). 'PBDEs in Italian Sewage Sludge and Environmental Risk of Using Sewage Sludge for Land Application'. *Environmental Pollution*. 161: 229–43; Vanguilder, C. (2012). *Hazardous Waste Management*. (Mercury, Boston). 172–87; Armendariz, C. (2011). 'Metals Monitoring in Sewage Sludge'. *Toxicology Letters*. 205: S195–S196; Anon. (2009). 'National Sewage Sludge Survey'. *BioCycle*. 50(3): 10, 12; Díaz-Cruz, M.S. (2009). 'Analysis of Selected Emerging Contaminants in Sewage Sludge'. *Trends in Analytical Chemistry*. 28(11): 1263–75; Singh, R. (2008). 'Potential Benefits and Risks of Land Application of Sewage Sludge'. *Waste Management*. 28(2): 347–58; Benidickson, J. (2007). *The Culture of Flushing: A Social and Legal History of Sewage*. (UBC Press, Vancouver). 91, 219–23; Harrison, E. (2006). 'Organic Chemicals in Sewage Sludges'. *Science of the Total Environment*. 367: 481–97; Amir, S. (2005). 'Sequential Extraction of Heavy Metals During Composting of Sewage Sludge'. *Chemosphere*. 59(6): 801–10.

## 3. NATIONAL AND INTERNATIONAL GOALS

Although these technical standards needed to adequately treat sewage have been evident since the turn of the 20th century, the application of these standards has taken much longer. For example, in the United States, the setting of standards for the treatment and disposal of sewage was not considered a matter of federal attention until 1934, when the Conservation of Wildlife, Fish and Game Act allowed the Secretary of Agriculture to investigate and determine 'the effects of domestic sewage …' and other polluting substances on wildlife.[13] Technical support to set standards and federal aid to build requisite plants to treat sewage only arrived in 1948 with the Water Pollution Control Act,[14] before being enhanced in 1956,[15] 1961[16] and 1966.[17] Finally, the monumental 1972 Clean Water Act[18] was passed. This Act was the first to declare, without reservation, that it was national policy that the discharge of toxic pollutants into the waterways was to be prohibited. To help in the achievement of this goal, federal financial assistance became available to construct publicly owned waste treatment works, based upon best practicable control technologies. In addition, public guidelines on water quality were to be propagated, with specific regard to acceptable levels of sewage and maximum daily thermal loads into waterways.[19] The success of this programme was such that by the 21st century, there were over 3,300 large treatment facilities operating under this regime, processing and treating more than 1 million gallons of sewage – per day.

The international community watched the United States and other developed countries replicate such practices, with resultant health benefits, multiple times. As the success of these campaigns became recognised, there was a concerted effort to replicate the results. This began when the discussion of global goals for sanitation arose in 1977 at the

---

[13] The 1934 Act on Conservation of Wild Life, Fish and Game. 48 Stat. 401, 1928–1934. Section 2. Also, Tarr, J. (1985). 'Historical Perspectives on Hazardous Wastes in the United States'. *Waste Management & Research*. 3: 95–102.

[14] The 1948 Water Pollution Control Act. 62 Stat. 115. 1918–1949.

[15] The 1956 Act to Extend and Strengthen the Water Pollution Control Act. 70 Stat. 498. 1956. Sections 2 and 6 in particular.

[16] The 1961 Act to Amend the Federal Water Pollution Control Act and to Provide for a More Effective Programme of Water Pollution Control. 75 Stat. 204. 1961.

[17] The 1966 Amendment to the Federal Water Pollution Control Act to Make it More Effective. 80 Stat. 1246. 1966.

[18] The 1972 Clean Water Act. 86 Stat. 816. 1972.

[19] Sections 101, 208, 301.

United Nations Water Conference. This was followed by the 1981 to 1990 International Drinking Water Supply and Sanitation Decade. This had the target of 'safe water and sanitation for everybody by 1990'. Two years later in 1992, the international community, conscious that 'faecal matter treatment and disposal should be accorded due priority given the potential threat of faeces to human health'[20] reset the target. Now, the goal was that by 1995, in industrialised countries, and by the year 2005, in developing countries, at least 50 percent of all sewage, wastewaters and solid wastes were treated or disposed of in conformity with national or international environmental and health quality guidelines. Thus, for both the developed and developing world, the goal was that 'by the year 2025, dispose of all sewage, wastewaters and solid wastes in conformity with national or international environmental quality guidelines'.[21] Ten years later, the Millennium Development Goals set the target to halve by 2015 the proportion of people who in 1990 were without basic access to sanitation. Two years later at the World Summit on Sustainable Development in 2002, the international community set an almost identical target to reduce by half the proportion of people without access to sanitation by 2015. This goal was again repeated with the International Water Decade (2005–2015). This Decade, linked to the 2008 International Year of Sanitation, reiterated the goal 'to halve, by 2015, the proportion of the world's population without sustainable access to basic sanitation'. Finally, in 2012, the international community at the Rio+20 Conference stressed 'the need to adopt measures to significantly reduce water pollution and increase water quality [and] significantly improve wastewater'.[22]

## 4. THE PRIMARY PROBLEM

As laudable as the international goals are, there has been limited progress in achieving them. As of 2014, 63 percent of the global population use improved sanitation facilities, as in a facility that at least separates human excrement from human contact. Over 2.5 billion people exist without

[20] Agenda 21 (1992). UNCED. Doc A/CONF.151/4. Chapter 27.

[21] Agenda 21 (1992). UNCED. Doc A/CONF.151/4. Chapter 21.29 and 21.39.

[22] United Nations, Report of the United Nations Conference on Sustainable Development, Rio de Janeiro, Brazil, 20–22 June 2012 at 24, paragraph 124. Adopted by the UN General Assembly. 66/288. *The Future We Want*. 27 July 2012.

improved sanitation, and 15 percent of the global population (some 1.1 billion people) practise open defecation in fields, forests, bushes, bodies of water or other open spaces. One-third of the 2.5 billion people without improved sanitation live in India, including some 626 million people who practise open defecation.[23] Whilst critics have been quick to point out that sanitation targets are unlikely to be achieved if the sanitation technologies used in industrialised countries (15,000 litres of water per person, per year and extensive infrastructure to remove the offensive material) are applied in low income countries, the fact remains that progress in this area – even using alternative technologies – has remained slow.[24]

## 5. THE SECONDARY PROBLEM

Once human waste has been taken out of the built environment, the secondary problem of what to do with it next arises. There are three possible answers to this. The first option, as discussed in Chapter 7, is to dispose of the waste into the ocean. However, as noted, this is a practice which a number of countries are increasingly phasing out. In addition, the international community, although not prohibiting the practice, is

---

[23] Kobel, D. (2012). 'Evaluation of Non-User Benefits Towards Improvement of Water and Sanitation Services in Informal Settlements'. *Urban Water Journal.* 9(5): 347–59; Isunju, D. (2011). 'Socio-Economic Aspects of Improved Sanitation in Slums'. *Public Health.* 125: 368–76; Konteh, F. (2009). 'Urban Sanitation and Health in the Developing World'. *Health & Place.* 15 (2009) 69–78; Cummings, O. (2009). 'The Sanitation Imperative'. *Desalination.* 248: 8–13; Greene, S. (2006). 'Water, Sanitation and Hygiene: Health 101'. *The Journal of the Royal Society for the Promotion of Health.* 126(2): 66–8; Ludwig, H. (2003). 'The Truth about Public Health Protection and Community Sewerage and Water Supply Sanitation in Developing Countries'. *The Environmentalist.* 23: 151–7.

[24] Schmidt, W. (2014). 'The Elusive Effect of Water and Sanitation on the Global Burden of Disease'. *Tropical Medicine and International Health.* 19(5): 522–30; Mazeau, A. (2014). 'Emerging Categories of Urban Shared Sanitation'. *Water and Environment Journal.* 28(4): 592–604; UNICEF/WHO (2012). *Progress on the Drinking Water and Sanitation.* (WHO, Geneva). 15–19; Henry, C. (2011). 'Fighting a Different Battle: Sanitation'. *BioCycle.* 52(8): 58–9; Murphy, H. (2009). 'Appropriate Technology: A Comprehensive Approach for Water and Sanitation in the Developing World'. *Technology in Society.* 31: 158–67; Black, M. (2008). *The Last Taboo: Opening the Door on the Global Sanitation Crisis.* (Earthscan, London). 48–67, 93–9, 103–8, 118–25, 132; Editor (2006). 'Municipal Solid Waste Management as a Luxury Item'. *Waste Management.* 26: 801–2.

attempting to regulate the disposal of sludge into the oceans to a much greater degree than previously.

In as much as many countries and the international community are trying to stop, if not regulate, the dumping of sewage and sludge into the ocean, the same groups are also trying to prevent the same waste entering river systems. The difficulty here is that pumping sewage into a river is often the easiest way to deal with it. This has been the practice for hundreds of years. However, since the end of the 19th century, it has become apparent that, depending on the type of sewage and how well it has been, or has not been, treated, this method can create a series of further environmental problems. The extent of the impact will depend upon the extent it is treated (and the extent that the viruses, heavy metals, organic pollutants and pharmaceuticals are removed), the amounts of waste involved, the biology of the river and the susceptibility of the associated ecosystems and species – all the way to the ocean where the rivers emerge and flow into.[25]

Due to such considerations, countries like Britain and the United States enacted legislation from the 1870s onwards, over how much sewage and how it could be disposed into freshwater systems. These laws have been progressively tightened over the following 140 years for, as successive generations have discovered, the safe and effective removal of pollutants from sewage waste which is a very difficult process, even in the most developed of countries.[26]

---

[25] Tamara, G. (2014). 'Season Variations and Resilience of Bacterial Communities in a Sewage Polluted Urban River'. *PLOS One*. 9(3): 456–62; Garcia, T. (2011). 'Antimicrobial Resistance of Heterotrophic Bacteria in Sewage-Contaminated Rivers'. *Water Research*. 45(2): 788–96; Rodríguez-Díaz, J. (2009). 'Detection and Characterization of Waterborne Gastroenteritis Viruses in Urban Sewage and Sewage-Polluted River Waters in Caracas, Venezuela'. *Applied and Environmental Microbiology*. 75(2): 387–94; Jacobson, H. (2008). 'Algal Growth Response in Rivers Receiving Sewage Effluent'. *Journal of Freshwater Ecology*. 23(2): 71–93; Williams, J. (2003). 'Steroid Estrogens Profiles along River Stretches Arising from Sewage Treatment Works Discharges'. *Environmental Science & Technology*. 37(9): 1744–50; Debruyen, Y. (2003). 'The Role of Sewage in a Large River Food Web'. *Canadian Journal of Fisheries and Aquatic Sciences*. 60(11): 1332–44; Ternes, T. (1998). 'Occurrence of Drugs in German Sewage Treatment Plants and Rivers'. *Water Research*. 32(11): 3245–60.

[26] The foremost British example in this area is the 1876 Rivers Pollution Prevention Act. 39 and 40 Victoria. See Parts I and II. This Act established general prohibitions against refuse, sewage and pollution from manufacturing, but exempted them if they employed the best practicable means of rendering the offending substances harmless into the waters. For commentary on the same

At the global level, when dealing with shared rivers, the broad obligation not to pollute rivers of international significance with all wastes, including sewage, dates back to the 1911 International Regulations Regarding the Use of International Watercourses for Purposes other than Navigation that were developed between the United States and Canada.[27] Soon after in 1918, the International Joint Commission Between the United States and Canada recognised untreated sewage as their foremost problem with their shared waterways,[28] of which a further four decades of talks were required before the problem began to be jointly mitigated via mutually agreed treatment and release standards of human waste before it would be allowed to enter shared waterways.[29]

At the global level, in terms of principles, it has been fairly clear since the 1970s that those who share rivers should not act to the unreasonable detriment of others.[30] The rule was finally spelt out in the 1997 Convention on the Law of the Non-Navigational Uses of International Watercourses which explained that all of the Parties were under an obligation to 'individually and, where appropriate, jointly, prevent, reduce and control the pollution of an international watercourse that may cause significant harm to other watercourse States or to their environment…'.[31] Specifically, States must exercise due diligence not to cause significant harm to the system that other countries rely upon. Within this context, States should seek to establish joint water quality objectives and

---

issue in the United States, see Tarr, J. (1985). 'Historical Perspectives on Hazardous Wastes in the United States'. *Waste Management & Research*. 3: 95–102.

[27] Institute of International Law. 24 *Annuaire de l'Institut de Droit International* (20 April 1911). See articles I and II(2).

[28] International Joint Commission (1918). *The Pollution of Boundary Waters*. (IJC, Atlantic City). 4–5.

[29] International Joint Commission (1950). *The Pollution of Boundary Waters*. (IJC, Atlantic City). 6–7.

[30] Bourne, C. (1996). 'The International Law Association's Contribution to International Water'. *Natural Resources Journal*. 36: 155–216 at 161; Recommendations I and II of the International Law Association. Report of the Forty-Ninth Conference, held in Hamburg, 8–12 August 1960 (London, 1961). xvi–xviii; and Report of the Fifty-Second Conference (Helsinki, 1967). 484–532.

[31] Article 21(2). Note, sewage is not explicitly mentioned, but can be easily inferred from the definition of pollution within the Convention: 'any detrimental alteration in the composition or quality of the waters of an international watercourse which results directly or indirectly from human conduct'. 1997 Convention on the Law of the Non-Navigational Uses of International Watercourses. Article 21(1).

pollution control techniques.[32] Although this Convention only entered into force in August of 2014, its broad principles had already been clearly articulated by the International Court of Justice in two separate cases, notably, in the *Pulp Mills River Case* of 2010, and the earlier *Case Concerning the Gabcíkovo-Nagymaros Project* of 1997.[33]

In many developed parts of the world, these principles, replete with strong standards to control sewage pollution of freshwater systems, exist. A foremost example of this is in Europe, via their overlapping Regional Convention on the Protection and Use of Trans-boundary Watercourses and International Lakes of 1992,[34] the Water Framework Directive of 2000[35] as well as specific treaties governing individual basins such as those for the Danube and the Rhine. Collectively, these agreements have set standards to cooperate for the conservation of fresh water, via standard setting and obligations to employ, as far as possible, the best available technologies to control pollutants. The Water Framework Directive complements these goals, by using a combined approach for the control of pollution by setting emission limit values for particular pollutants and water quality objectives, against an ecological context of the needs of each freshwater system.[36] Supplementary Directives, such as that Concerning the Management of Bathing Water Quality, focus the targets in this area by the public monitoring of the bathing quality of beaches in Europe by concentrating upon two parameters (intestinal enterococci and *Escherichia coli*) that are directly related to sewage and human bodily waste.[37]

---

[32] See articles 5, for equitable use, and 7, for significant harm, and 21(1), (2) and (3). For commentary, see McCaffrey, S. (2007). *The Law of International Watercourses*. (OUP, Oxford). 439–40.

[33] *Pulp Mills River*. Judgment. I.C.J. Reports 1997. Note, this favourably quotes the *Gabcíkovo-Nagymaros Project (Hungary/Slovakia)* case on similar principles. See paragraphs 140, 181–9, 195–7 and 204; Plakokefalos, I. (2011). 'The Pulp Mills Case'. *The International Journal of Marine and Coastal Law*. 26: 169–83.

[34] Also known as 'the Helsinki Convention', was adopted at Helsinki under the auspices of the UN Economic Commission for Europe on 17 Mar. 1992 and entered into force on 6 Oct. 1996. 31 ILM 1312.

[35] The EU Water Framework Directive 2000/60/EC (23 Oct. 2000).

[36] See the preamble, articles 2, 8 and 10 and Annexes 4 and 5 of the Water Framework Directive, ibid. Also, see Chave, P. (2001). *The EU Water Framework Directive: An Introduction*. (EC, Brussels). 7–10.

[37] See Directive 2006/7/EC (15 Feb. 2006) Concerning the Management of Bathing Water Quality.

Whilst clear progress is being made in such high income countries sharing freshwater systems, the same is not obvious in other parts of the developing world. This is evident on two counts. First, many areas do not even have cooperative agreements in place to regulate the pollution of these systems, let alone in specific terms with regard to sewage. Second, if they do, the achievements are often modest or uneven in their geographical coverage or standards of control, as is the case with China. In other instances, such as with India, the percentage of sewage which is adequately treated may be as low as 20 percent. Correspondingly, 80 percent of the pollution load destroying the rivers in India is untreated human waste.[38]

When the countries in the developing world have to share their rivers, this problem can become even more difficult to resolve, despite a number of bilateral treaties attempting to regulate the problem. For example, with the Indus, despite commitments by both Parties in their Waters Treaty of 1960 to take all reasonable measures to ensure that untreated sewage did not enter the waterway, the practice goes on largely unabated. In other parts of the world such as the Jordan river basin, which is currently under threat due to some 13 million cubic metres of effluent entering the waterway each year, whilst the commitment by many of the interested parties (Israel, Jordan and the Palestinian Liberation Organization) to control and treat sewage before it is discharged into the waterways appears on paper, the willingness to implement these goals in a highly charged environment is sometimes questionable.[39]

---

[38] Jin, L. (2014). 'Current State of Sewage Treatment in China'. *Water Research*. 66: 85–98; Porter, C. (2013). 'Sewage Pollution of the River Ganga'. *River Systems*. 20(3): 303–27; Suwarno, D. (2014). 'Fast Increases in Urban Sewage Inputs to Rivers of Indonesia'. *Environment, Development and Sustainability*. 16(5): 1077–97; Xia, Y. (2013). 'Diurnal Pattern in Nitrous Oxide Emissions from a Sewage Enriched River'. *Chemosphere*. 92(3): 303–24.

[39] Miner, M. (2009). 'Water Sharing Between India and Pakistan: A Critical Evaluation'. *Water International*. 34(2): 204–16; Black, M. (2008). *The Last Taboo: Opening the Door on the Global Sanitation Crisis*. (Earthscan, London). 6; Masood, S. (2005). 'Hazardous Effects of Sewage Water on the Environment'. *Management of Environmental Quality: An International Journal*. 16(4): 338–52; Rifatta, W. (2004). 'Water Quality Assessment of the River Nile System: An Overview'. *Biomedical and Environmental Sciences*. 17: 87–100; Wiebe, K. (2001). 'The Nile River: Potential for Conflict and Cooperation in the Face of Water Degradation'. *Natural Resources Journal*. 41(3): 731–54; Shamir, U. (1997). 'Water Agreements between Israel and Its Neighbors'. *Middle Eastern Natural Environments Bulletin*. 103: 274–96; Concannon, B. (1989). 'The Indus

If sewage-related waste cannot be safely or legally disposed of into the ocean or rivers, the only options left are to deal with it on land, via either landfilling, incinerating or recycling it. Each option has potential benefits, and potential risks. At the moment, there is no dominant practice, with different regions pursuing alternative options.[40]

The incineration of dried sludge is becoming increasingly common in countries which have a shortage of space to landfill. This is evident with countries like Japan burning 70 percent of their sludge whilst over 58 percent is incinerated in the Netherlands. The difficulty with the incineration of sludge is the ash residue, which is often heavily contaminated with residues such as heavy metals that make the disposal of the hazardous ash problematic. Although some safe uses for this ash are possible, such as being turned into certain building materials, such transformations are rare.[41]

The second approach is to use dried sewage sludge as a type of cover to be placed in, and on, landfill. This practice is common in both China and large parts of Europe. In all instances, the environmental success of the action depends upon how well the sludge has been dried and neutralised first, and secondly, the adequacy of the landfill site to deal with the resultant leachate.[42]

---

Waters Treaty: Three Decades of Success?'. *Geographical International Environmental Law Review*. 2: 55–80.

[40] Yoshida, H. (2013). 'Life Cycle Assessment of Sewage Sludge Management'. *Waste Management and Research*. 31(11): 1083–110.

[41] Cieslik, B. (2014). 'Review of Sewage Sludge Management: Standards, Regulations and Analytical Methods'. *Journal of Cleaner Production*. 11: 401–20; Werle, S. (2014). 'Analysis of Organic and Inorganic Contaminants in Dried Sewage Sludge and By-Products of Dried Sewage Sludge Gasification'. *Energies*. 7(1): 462–80; Pavsic, P. (2014). 'Sewage Sludge Ash Based Products for Sustainable Construction'. *Journal of Cleaner Production*. 67: 117–24; Lin, H. (2012). 'Simulation of Co-Incineration of Sewage Sludge with Municipal Solid Waste'. *Waste Management*. 32(3): 561–7; Chen, T. (2012). 'Fixation and Partitioning of Heavy Metals in Slag after Incineration of Sewage Sludge'. *Waste Management*. 32(5): 957–64; Roy, M. (2011). 'Review of Bio-Solids Management Options and Co-Incineration'. *Waste Management*. 31(11): 2228–35; UN-Habitat (2008). *Global Atlas of Excreta, Wastewater Sludge, and Biosolids* Management. (Habitat, Nairobi). 27–9; Choi, C. (2004). 'Power Sludge'. *Scientific American*. May. 20.

[42] Chen, H. (2012). 'Utilisation of Urban Sewage Sludge: Chinese Perspectives'. *Environmental Science and Pollution Research*. 19(5): 1454–68; Hyun, J. (2012). 'Conversion of Sewage Sludge to Daily Landfill Cover Material'. *Journal of Material Cycles and Waste Management*. 14(1): 14–24; Song, U. (2010). 'Environmental and Economical Assessment of Sewage Sludge Compost

The third approach is where the human waste is recycled into what is commonly known as 'biosolids' and fed back into a biotic system. The basis of this equation is that human excreta includes diverse and complex molecules derived from foods and bodily processes – such as carbohydrates, sugars and fats. These molecules include common nutrients, such as nitrogen and phosphorus, as well as micronutrients, such as copper, iron, nickel and zinc, that, if within correct amounts, are useful for the growth of many types of biotic matter. The organic matter in biosolids is also of value to natural processes and chemical bonds, boosting the soil's organic matter content and nutrient reserves. The cumulative result is that, in theory, an average human's production of waste (500 litres of urine and 50 kilograms of faeces per year) could lead to 200 kilograms of cereals being grown and fertilised.[43]

This is not a new realisation. All civilisations which developed cesspits had to have a way to clean them out. The most common method of all was with the use of 'night-soil men' who were employed during the hours of darkness to dig up the foul smelling contents of cesspools. These contents were carried away and dried out in holding areas, then ferried to the countryside, often combined with horse dung, where it was used as fertiliser. Successful use of this type of waste can be traced to the 15th century. The process continued until large scale sewage systems, flushed with water, carried the waste further afield. Here, although there were a number of attempts at large scale sewage farming with treated waste, the successes did not trump the economic costs, nor social misconceptions, involved.[44]

In the 21st century, a few countries are trying to reverse this situation, and make a more active use of biosolids once they have been adequately

in a Landfill'. *Resources, Conservation and Recycling*. 54(12): 1109–20; O'Kelly, B. (2005). 'Sewage Sludge to Landfill'. *Journal of the Air and Waste Management Association*. 55(6): 765–80.

[43] Rizzardini, C. (2014). 'Sustainability of Domestic Sewage Sludge Disposal'. *Sustainability*. 6(5): 2424–40; Kruger, O. (2014). 'Complete Survey of German Sewage Sludge Ash'. *Environment, Science and Technology*. 48(20): 11811–12825; UN-Habitat (2008). *Global Atlas of Excreta, Wastewater Sludge, and Biosolids Management*. (Habitat, Nairobi). 22–9.

[44] Benidickson, J. (2007). *The Culture of Flushing: A Social and Legal History of Sewage*. (UBC Press, Vancouver). 120–27; Black, M. (2008). *The Last Taboo: Opening the Door on the Global Sanitation Crisis*. (Earthscan, London). 30–31; Simmons, D. (2006). 'Waste Not, Want Not: Excrement and Economy in Nineteenth Century France'. *Representations*. 96: 73–98; Hamlin, A. (1980). 'Sewage: Waste or Resource – A Historical Perspective'. *Environment*. 22(8): 16–34.

treated and stripped of possible pollutants. However, this is slow work, as regulators and scientists are attempting to balance out the possible positive, and negative, benefits of the application of biosolids. The key considerations involve levels of treatment required for the biosolids, when and how they may be applied, and most importantly, how much may be applied and on what lands. The cultural acceptability of this practice is also important. Through such practices, 81 percent of all biosolids are recycled to the soils in Australia, 66 percent in New Zealand and 55 percent in the United States. In these countries, the biosolid products are developing an economic value, sometimes up to a few hundred dollars per ton, depending on how they have been treated.[45]

## 6. CONCLUSION

Human waste is the most personal, and lethal, of all waste streams. It is also the first type of waste that humanity began to manage. Over the following 6,000 years, as the population and risks of disease have increased, so too have the scientific understanding and technical ability to move the waste out of places of human congestion. In the last century, the first generation of problems involved getting the waste out of the cities. Second generation problems involved getting the waste suitably treated and neutralised to the greatest possible degree, and then finally, as

[45] For the European rules in this area, see Council Directive 86/278/EEC (12 June 1986) On Protection of the Environment, and in Particular of the Soil, when Sewage Sludge is Used in Agriculture. In particular, articles 5 to 10. The quantities are set out in Annex I.A, I.B and I.C. For commentary, Synder, E. (2013). 'Risk Assessment of Land Applied Biosolids'. *The Science of the Total Environment*. 442: 437–50; Pearce, F. (2013). 'Flushed with Success'. *New Scientist*. 16 Feb. 49; Hodson, H. (2012). 'Waste Not, Want Not'. *New Scientist*. 1 Sept. 17; Goven, J. (2012). 'Community Engagement in the Management of Biosolids: Lessons from New Zealand'. *Journal of Environmental Management*. 103: 154–66; Lu, Q. (2012). 'Land Application of Biosolids in the USA'. *Applied and Environmental Soil Science*. 2012: 1687–99; McFarland, M. (2012). 'Groundwater Quality Protection at Biosolid Land Application Sites'. *Water Research*. 46(18): 5963–75; Brisolara, K. (2010). 'Biosolids and Sludge Management'. *Water Environment Research*. 82(10): 1311–20; UN-Habitat (2008). *Global Atlas of Excreta, Wastewater Sludge, and Biosolids Management*. (Habitat, Nairobi). 22–9; Rogers, H. (2005). *Gone Tomorrow: The Hidden Life of Garbage*. (Norton, NYC). 34–5; Bright, D. (2003). 'Contaminant Risks from Biosolids Land Application'. *Environmental Pollution*. 126(1): 39–50; Renner, R. (2000). 'Sewage: Pros and Cons'. *Environmental Science and Technology*. Oct. 25–40.

a third generation problem, effectively and safely disposing of the residual waste, which is typically in the form of sludge.

This need, to remove human waste safely from places where it can cause disease, is a well-recognised international goal. However, success in meeting this goal is not the experience of all countries. Whilst the overwhelming majority of developed countries can deal adequately with this problem, a number of developing countries are still struggling with the first parts of this equation, trying to ensure adequate sanitation and basic evacuation procedures for this waste.

It is only for the countries which have mastered the first generation question of moving the human waste away from settlements that the issues of the final disposal of human waste come to the fore. The options for final disposal are into freshwater systems and/or the ocean; or disposal via either incineration, placement into landfill or recycling. Regarding the water option, whilst the international community has come to regulate, not prohibit, the dumping of such wastes into the ocean with a relative degree of success, the same cannot be said for freshwater systems. Here, despite some broad policy goals at the global level, the success of keeping human waste out of freshwater systems has ranged from very good in many developed countries, to very poor in many developing countries. In respect of the other options of disposal via incineration, landfill or recycling, each approach has its benefits, and its limits. It is possible that the last option, where biosolids are fed back into a well-managed biotic system, is the most advantageous use of the resource. However, this is only utilised by a few countries, and even then, within a tightly regulated context.

# 9. Nuclear waste

## 1. INTRODUCTION

The last chapter in this book is about nuclear waste. The reason that this waste stream has a specifically dedicated chapter is because this is the most hazardous of all waste types. Ironically, when the management of nuclear considerations was first legislated for in a domestic setting with the 1946 Atomic Energy Act of the United States, the topic of waste was not in the remit of consideration. However, within a dozen years, this was the very the first type of waste to be specifically focused upon by international law. By the end of the century, once it had become apparent that this was likely to be the last waste left on the planet, tens of thousands of years from now, it became a dominant international concern. Despite this concern, and over four decades of debate, the only international rule that is certain is that it should not be dumped into the ocean. Rather, it should be reprocessed to minimise the amount and/or then set into deep geological storage. There are, however, significant problems with both of these options.

## 2. REPROCESSING AND/OR FINAL DISPOSAL OF NUCLEAR WASTE

The recycling in the context of nuclear material is known as *reprocessing*. Reprocessing is a term largely restricted to nuclear matters, meaning a process or operation, the purpose of which is to extract radioactive isotopes from spent fuel for further use.[1] This involves chemically separating uranium and plutonium from spent nuclear fuel so that they can be re-enriched (the uranium) or refabricated (the plutonium) and reused in specialised reactors. A number of countries are currently engaged in reprocessing. Major plants operate in France, the United

[1] Article 2(l). 1997 Joint Convention on the Safety of Spent Fuel Management and on the Safety of Radioactive Waste Management. IAEA. INFCIRC/546. 24 Dec. 1997.

Kingdom and Russia. Belgium, China, India and Switzerland are also active in this space. Since 1970, over 80,000 tons of used nuclear fuel has been recycled. France, Britain and Russia undertake reprocessing for their own nuclear industries and those located in other countries. Most notably, Japan has made over 140 shipments of used nuclear fuel to Europe for reprocessing since 1979, with the vitrified waste and the recovered uranium and plutonium (as MOX – mixed oxide – fuel) being returned to Japan to be used in fresh fuel.[2]

The strong interest in reprocessing is because it serves multiple purposes. Originally reprocessing was used solely to extract plutonium for producing nuclear weapons. With the commercialisation of nuclear power, the practice developed whereby reprocessed plutonium could be extracted from used fuel and recycled. As it stands, only about 5 percent of the value of nuclear fuel is extracted when taken 'once through'. This figure can be changed if the used fuel is converted through a MOX fuel fabrication plant where it is mixed with depleted uranium oxide to make fresh fuel for thermal reactors, whereby some 25 to 30 percent more energy can be extracted from the original uranium.[3] In addition, reprocessing reduces the size of the high-level waste (by about 80 percent) that has to be dealt with. This is an important point, in that even if large scale reprocessing was undertaken, a residual amount of nuclear waste will still require long term storage.[4]

The answer to what is the best type of long term storage for nuclear waste on land depends upon what type of waste it is. Although low, medium and high grade forms of nuclear waste require engineered surface structures of tightly isolated containment cells, the answer to what is the best type of storage for nuclear waste on land depends upon what type of waste it is. For low-level waste, the general goal is that it must be held for at least 300 years before it becomes safe. With the medium level, the goal is to place it in a position in which it can be deposited with higher level types of waste when possible. Globally, there

---

2 Silverio, L. (2011). 'An Analysis of Development and Research on Spent Nuclear Fuel Reprocessing'. *Energy Policy*. 39: 281–9; Poluektov, P. (2010). 'Moscow's Liquid Radwaste Reprocessing Plant'. *Atomic Energy*. 108(2): 127–30.

3 Zagar, T. (2011). 'Recycling as an Option of Used Nuclear Fuel Management Strategy'. *Nuclear Engineering and Design*. 241: 1238–42; Choppin, G. (2006). 'Technology for Nuclear Reprocessing: Present and Future Directions'. *Separation Science and Technology*. 41: 1955–63.

4 Hippel, F. (2001). 'Plutonium and Reprocessing of Spent Nuclear Fuel'. *Science*. 293(28): 2397.

are about 40 near surface large scale disposal facilities for the taking of low-level and medium-level nuclear waste. An additional 30 facilities are expected to be in operation by 2030. In time, the goal for many is that all of this waste, currently on the surface of the land, will be taken down into deep underground sites.

In this area, after over 50 years of debate, in the language of the European Community 'it is broadly accepted at the technical level that, at this time, deep geological disposal represents the safest and most sustainable option as the end point of the management of high-level waste and spent fuel considered as waste'.[5] At the moment, the current practice is that the overwhelming majority, about 270,000 tons of used fuel, is in such surface storage. About 90 percent of this is in storage ponds (which shield and cool the waste) of reinforced concrete with steel liners. The balance is in dry storage. After some years, the fuel can be removed and put into dry cask storage. In the United States, there are more than 800 waste-containing concrete casks currently on the open surface at 34 sites. In the future, the hope is that all such waste will eventually end up in deep geological storage where it will be safely entombed in perpetuity.[6]

The aim of deep geological storage is to place the waste in an environment where there will be no need for custodial care of the site after emplacement and closure of the access portals. This is not to suggest that this will be a perfect method for disposal of such waste, only that it is considered better than any of the alternatives (such as space-dumping of solid waste or drilling deep holes 5 kilometres into the crust of the Earth, and filling them with liquid waste). The basis of this thinking is that the placement of the high-level nuclear waste, deep underground, within robust geological formations should be relatively safe from a multitude of risks, ranging from terrorist attacks through to accidents, as was amply demonstrated with the Fukushima crisis in Japan in 2011 and the dangers of storing spent fuel in pools on the surface next to the ocean.[7]

---

[5] Paragraph 23 of the preamble to Directive 2011/70/EURATOM (19 July) on a Community Framework for the Responsible and Safe Management of Spent Fuel and Radioactive Waste.

[6] Hannum, J. (2005). 'Smarter Use of Nuclear Waste'. *Scientific American*. Dec. 12–18.

[7] Yokoyama, J. (2013). 'Fukushima Disaster and Reform'. *Environmental Policy and Law*. 43(4): 76–84; Alley, W. (2011). *Too Hot to Touch: The Problem of High Level Nuclear Waste*. (Cambridge UP, Cambridge). 112–21; Vandenbosch, R. (2007). *Nuclear Waste Stalemate*. (Utah UP, Salt Lake). 8–9.

The deep sites are usually characterised by chemically reducing conditions – that is, the absence of oxidants like dissolved oxygen which could encourage barrier corrosion or radionuclide dissolution. The host rock is meant to be seismically stable, with a low speed of vertical movements, where there is no volcanism or other natural processes, such as groundwater, that would allow the waste to leach. The potential rock formations may be classified as either wet or dry. Dry rocks (such as salt or anhydrite) do not contain a groundwater flow system and have no direct mechanism for degradation of the engineered barriers or transport of radionuclides. Wet rocks do contain mobile groundwater but may also be suitable if groundwater flow is low and is either localised in well-defined zones (such as crystalline rocks) or occurs only by very slow diffusion throughout the entire rock (such as clays). The corrosion resistant, very long term storage containers are then meant to be packed with impermeable backfill such as bentonite clay with additional engineered barriers and associated safety standards. Collectively, these protected sites are meant to ensure the inviolability of the waste in perpetuity (or at least a million years).[8]

## 3. THE INTERNATIONAL STANDARDS

International law recognises that the choice of how to deal with nuclear waste rests with each state and that whilst some states consider spent fuel as a valuable resource that should be reprocessed, others take an alternative view.[9] In addition, although the trade in nuclear waste for

---

8 Grarnbow, B. (2014). 'Geological Disposal of Nuclear Waste'. *Applied Geochemistry*. 49: 247–58; Alley, W. (2011). *Too Hot to Touch: The Problem of High Level Nuclear Waste*. (Cambridge UP, Cambridge). 10, 23–7; Laverov, N. (2010). 'Crystalline Rocks as a Medium for Nuclear Waste Disposal'. *Russian Journal of General Chemistry*. 81(9): 1980–93; Johnson, O. (2010). 'High-Level Radioactive Waste Management'. *Water Environment Research*. 82(10): 1748–1756; Anon. (1997). 'A History of High-Level Nuclear Waste Disposal Efforts'. *Environment*. 39(3): 10; Bird, G. (1982). 'The Nuclear Waste Disposal Problem: An Overview from a Geological and Geochemical Perspective'. *Chemical Geology*. 36: 1–13; McKinley, I. (1987). *Geological Disposal of Nuclear Waste*. (Wiley, NYC).

9 Preamble. The 1997 Joint Convention on the Safety of Spent Fuel Management and on the Safety of Radioactive Waste Management INFCIRC/546. (24 December 1997).

disposal has been prohibited in Antarctica,[10] Africa[11] and the South Pacific,[12] there is no generic international prohibition on the trade in used nuclear material.[13] In the case of trading waste for both reprocessing and/or final disposal, international law is receptive to such exchanges, although the idealised option is that 'radioactive waste should, as far as is compatible with the safety of the management of such material, be disposed of in the State in which it was generated'.[14] However, 'in certain circumstances, safe and efficient management of spent fuel and radioactive waste might be fostered through agreements among Contracting Parties to use facilities in one of them for the benefit of the other'.[15]

Accordingly, whilst it is recognised that 'any State has the right to ban import into its territory of foreign spent fuel and radioactive waste'[16] the flip side is also recognised that a state has the, prima facie, right to import such material as either waste or reprocessing stock if it desires to do so. Such importations are only permissible if prior notification of the shipment and clear consent of the state of destination is obtained. In addition, a state can only accept the transboundary movement of nuclear waste if it has the administrative and technical capacity, as well as the

---

10 See article 5 of the Antarctic Treaty. 402 UNTS. 71. Note, the 1991 Protocol on Environmental Protection to the Antarctic Treaty. BH992.txt, Annex III, Article 2.

11 See Article 2 of the 1991 Bamako Convention on the Prohibition of the Importation of Hazardous Waste into Africa, which explicitly included radiological waste as a type of prohibited hazard.

12 Article 4(1)(a). The (Waigani) Convention to Ban the Importation into Forum Island Countries of Hazardous and Radioactive Wastes and to Control the Transboundary Movement and Management of Hazardous Wastes within the South Pacific Region

13 Holland, I. (2010). 'Waste Not Want Not?'. *Australian Journal of Political Science*. 37(2): 283–301; MacFarlane, A. (2006). 'Stuck on a Solution'. *Bulletin of Atomic Scientists*. May. 46–52; Anon. (2005). 'Nuclear Suit'. *New Scientist*. 17 Dec. 5; Edwards, R. (2004). 'Fury at UK's Plans to Ship Hot Waste out to Kyrgyzstan'. *New Scientist*. 25 Sep. 15; Anon. (2004). 'Nuclear Waste? No Thanks'. *New Scientist*. 16 Oct. 5; Anon. (2004). 'Tibet Is Nuclear Dump'. *New Scientist*. 9 Oct. 5; Webster, P. (2002). 'The Grab for Trash'. *Bulletin of Atomic Scientists*. Sep. 33–7, 40–45; Glasser, S. (2001). 'Russian Seek Windfall from Nuclear Waste'. *New Scientist*. Feb. 22–8, 29; Guinnessy, P. (2001). 'Russia Banks on Importing Nuclear Waste'. *Physics Today*. 54(9): 24.

14 Paragraph 11 of the preamble to the Joint Convention on the Safety of Spent Fuel Management and on the Safety of Radioactive Waste Management.

15 Preamble. 1997 Joint Convention on the Safety of Spent Fuel Management and on the Safety of Radioactive Waste Management.

16 See paragraph xii of the preamble.

regulatory structure, needed to manage the spent fuel or the waste in a manner consistent with the 1997 Joint Convention on the Safety of Spent Fuel Management and on the Safety of Radioactive Waste Management.[17] Moreover, the state of origin must satisfy itself that any such assertion that they can adequately deal with the waste or reprocessing is correct. If this cannot be assured, then the material must not be exported.[18] Some regions, such as Europe, have taken this option of other countries importing their nuclear waste to only be acceptable if the importing country explicitly meets safety standards approved by the European Community.[19]

The defining international instrument in this area is the 1997 Joint Convention on the Safety of Spent Fuel Management and on the Safety of Radioactive Waste Management (the Joint Convention).[20] This Convention seeks to achieve and maintain a high level of safety worldwide in radioactive waste management. Fundamentally, it aims to ensure that during all stages of radioactive waste management there are effective defences against potential hazards so that individuals, society and the environment are protected. In doing so, the needs and aspirations of the present generation are meant to be met without compromising the ability of future generations to meet their needs and aspirations.[21]

Under the Joint Convention, the starting point for any reprocessing or waste disposal is that:

> Effective protection of individuals, society [including being cognisant of future generations] and the environment must be achieved by applying suitable protective methods as approved by the regulatory body, in the

---

[17] IAEA. INFCIRC/546. 24 Dec. 1997.

[18] Article 27.

[19] See article 4 of Directive 2011/70/EURATOM (19 July) on a Community Framework for the Responsible and Safe Management of Spent Fuel and Radioactive Waste.

[20] Note the very important supplemental work of the International Atomic Energy Agency (IAEA), the Nuclear Energy Agency of the Organisation for Economic Co-operation and Development, the European Commission and the International Commission on Radiological Protection (ICRP). See generally, IAEA (2012). *Measures to Strengthen International Cooperation in Nuclear, Radiation, Transport and Waste Safety.* GOV/2012/28-GC(56)/6. The most pertinent guidelines in this area are the IAEA (1995). *Principles of Radioactive Waste Management.* (IAEA, Vienna).

[21] Article 1.

> framework of its national legislation which has due regard to internationally endorsed criteria and standards.[22]

This point is very important, for as with all nuclear-related matters, national legislation is the first port of call. However, this port should be linked to international standards and guidelines, for which the Joint Convention provides the framework. Of this international framework, two sets of standards are relevant. The first set relate to the facility. Specifically, each nuclear recycling and/or waste facility has to be licensed and strictly regulated. All radiation exposure must be kept as low as is reasonably achievable. The facility must be prepared for emergencies, possess suitable and adequate staff and financial resources to function effectively, and if the operator cannot maintain the safety, the state is obliged to take responsibility.[23]

The second set of standards relates to the sites. Existing sites must be reviewed for safety standards (and all practicable improvements for safety are to be made), and proposed sites must be evaluated for all factors 'likely to affect the safety of such a facility during its operating lifetime'.[24] With proposed facilities, sites must be evaluated in light of 'site-related factors [including those which may evolve] likely to affect the safety of such a facility during its operating lifetime as well as that of a disposal facility after closure'.[25] If a breach occurs after the site is closed, it is expected that the unplanned release of radioactive material will be dealt with.[26] The design and construction of proposed facilities should be based on established technologies and provide for suitable measures to limit possible radiological impacts including those that are unplanned. Systematic safety assessments should be undertaken, for which the facility should be regularly maintained, monitored and inspected.[27]

---

22 Article 4(iv). Note also overarching article 1. The reference to future generations comes from article 4(vi) and (vii).

23 Articles 18, 19, 20, 21, 22, 24 and 25.

24 Article 6.

25 Article 13(1).

26 See articles 14(1), 12, 15 and 17.

27 Articles 5, 7, 8 and 9.

## 4. THE PROBLEMS IN PRACTICE

Although international law allows, and a number of states practise, either reprocessing and/or the active exploration of long term geological storage options, both considerations are subject to substantial difficulties.

With regard to nuclear reprocessing, a number of countries such as Germany, Canada, Finland, Spain, Sweden, South Korea and the United States have rejected this approach. There have been three reasons for this. First, the challenge of getting adequate safety standards for the reprocessing sites has proven difficult in some countries. Second, nuclear reprocessing has been linked to nuclear proliferation, as was clearly developed with India and its nuclear arsenal in the 1970s. In the new century, the same fear exists that uncontrolled reprocessing can lead to a direct supply of weapons grade plutonium. Third, the economic costs of reprocessing are often higher than the once-through fuel cycle due to the cheaper prices of uranium ore. These economic costs become even more prohibitive if additional technologies are utilised to prevent the risks of proliferation. Finally, as noted above, even if reprocessing is undertaken, a significant amount of waste still has to be permanently disposed of. As such, reprocessing may be seen as a distraction from the larger issues, of the generation of the waste, in the first place, and the disposal of the waste, in the last place.[28]

---

[28] Serfontein, D. (2014). 'Thorium Based Fuel Cycles: Reassessment of Fuel Economics and Proliferation Risk'. *Nuclear Engineering and Design*. 271: 106–13; Soria, B. (2013). 'Recycling versus Long-Term Storage of Nuclear Fuel: Economic Factors'. *Science and Technology of Nuclear Installations*. 13: 1–7; Rechtenwald, G. (2012). 'Cost Probability Analysis of Reprocessing Spent Nuclear Fuel in the US'. *Energy Economics*. 34: 1873–81; Widder, S. (2010). 'Benefits and Concerns of a Closed Nuclear Fuel Cycle'. *Journal of Renewable and Sustainable Energy*. 2(6): 124–37; Beken, T. (2010). 'Security Risks in Nuclear Waste Management'. *Journal of Environmental Management*. 91(4): 940–48; Diyakov, A. (2010). 'The Nuclear Renaissance and Preventing the Spread of Enrichment and Reprocessing Technologies'. *Daedalus*. 139(1): 117–25, 138; Spring, N. (2009). 'Rethinking Nuclear Recycling'. *Power Engineering*. 113(7): 94–6; Anon. (2009). 'Adieu to Nuclear Power'. *Nature*. 460: 152; Schneider, E. (2009). 'Cost Analysis of the US Spent Nuclear Fuel Reprocessing Facility'. *Energy Economics*. 31: 627–34; Lindemyer, J. (2009). 'The Global Nuclear Energy Partnership: Proliferation Concerns and Implications'. *The Nonproliferation Review*. 16(1): 79–93; Hippel, J. (2008). 'Rethinking Nuclear Fuel Recycling'. *Scientific American*. May. 88–96; Vandenbosch, R. (2007). *Nuclear Waste Stalemate*. (Utah UP, Salt Lake). 16–18, 50–60; Dawson, J. (2005). 'Proliferation is Key Issue in Nuclear Power Resurgence'. *Physics Today*. 58(7): 33; Hippel, F. (2001). 'Plutonium and Reprocessing of Spent Nuclear

In terms of the final disposal of low and medium grade nuclear wastes, despite the relative clarity on the applicable technologies and principles needed to achieve the safe management of this waste over the last 40 years, the handling of this waste has seen great difficulties in many countries. This has been especially related to the containers used to hold such waste, with large questions of their adequacy over a period of decades, let alone centuries.[29]

The same concern exists for questions about the adequacy of storage containers for high grade nuclear waste. This high grade waste is typically vitrified into borosilicate/pyrex glass, encapsulated into heavy stainless steel cylinders about 1.3 metres high before being taken to permanent disposal deep underground. The difference from low and medium grade waste is that the storage containers for this must last into the indefinite future. There are serious questions over whether it is possible for such containers to maintain their integrity for the following millennia.[30]

It is the problem of perpetuity that has caused many scholars and policy makers to question this method of disposal. The basis of the questioning is whether it is reasonable to assume that the method is safe when projected out over thousands of years, let alone hundreds of thousands of years. In this time, everything from technological failure through to risks from human action (accidents, war, mistake) through to

---

Fuel'. *Science*. 293 (5539): 2397–405; Edwards, R. (2000). 'Stash or Burn?' *New Scientist*. 11 Nov. 5.

[29] Alley, W. (2011). *Too Hot to Touch: The Problem of High Level Nuclear Waste*. (Cambridge UP, Cambridge). 131–51; Duffo, G. (2010). 'Corrosion Susceptibility of Steel Drums Containing Cemented Intermediate Level Nuclear Wastes'. *Journal of Nuclear Materials*. 405: 274–9; Gallagher, E. (2009). 'Overview of Design of Modular Vaults for the Storage of Low Level Nuclear Waste'. *Materials Research Society Proceedings*. 1193: 50–62; Anon. (2005). 'Overrun by Waste'. *New Scientist*. 10 Dec. 7; Gudowski, W. (2005). 'Nuclear Waste Management'. *Nuclear Physics, Section A*. 752: 623–35; Carpenter, T. (2004). 'Don't Breathe the Air'. *Bulletin of Atomic Scientists*. May. 24–8; Edwards, R. (2000). 'Nuclear Muddle'. *New Scientist*. 9 Sept. 14; Edwards, R. (1999). 'Trouble in Store'. *New Scientist*. 2 Jan. 7; Lacoste, A. (1997). 'The French Approach to the Safe Management of Very Low Level Radioactive Waste from Nuclear Installations'. *Nuclear Engineering and Design*. 176: 151–3; Anon. (1996). 'Hot Secrets'. *New Scientist*. 6 Jan. 12; Newman, A. (1995). 'Low Level Radioactive Wastes: High Level Risk'. *Environmental Science and Technology*. 28(11): 485–95.

[30] King, F. (2014). 'Predicting the Lifetimes of Nuclear Waste Containers'. JOM. 66(3): 526–35; Hippel, F. (2001). 'Plutonium and Reprocessing of Spent Nuclear Fuel'. *Science*. 293(28): 2397.

slow processes (from climate change to tectonic movements) and fast natural movements (such as earthquakes, volcanism, groundwater change) could directly impact upon the safety of the disposal method.[31]

It was due to such problems that countries such as Britain and Germany changed their positions back and forth over the suitability of deep geological disposal. In Japan, following the Fukushima crisis, there is great uncertainty about whether to endorse deep geological disposal or not. The United States joined this uncertainty in 2012, when their project of deep geological disposal of high-level waste in the Yucca Mountain complex was cancelled, after a process begun 30 years earlier with their 1982 Nuclear Waste Policy Act.[32] Conversely, in Finland, France and Sweden, deep geological disposal facilities have been approved and are expected to start operating after 2025.[33]

---

[31] Naslund, J. (2013). 'Climate Considerations in Long-Term Safety Assessments for Nuclear Waste Repositories'. *AMBIO*. 42: 393–401; Beken, T. (2010). 'Security Risks in Nuclear Waste Management'. *Journal of Environmental Management*. 91: 940–48; Rutqvist, J. (2009). 'A Multiple-Code Simulation Study of the Long-Term EDZ Evolution of Geological Nuclear Waste Repositories'. *Environment and Geology*. 57: 1313–24; Cattant, M. (2008). 'Corrosion Issues in Nuclear Industry Today'. *Materials Today*. 11(10): 30–42; Feron, D. (2008). 'Corrosion Issues in Nuclear Waste Disposal'. *Journal of Nuclear Materials*. 379: 16–23; Louberg, H. (2002). 'Long-Term Risk Management of Nuclear Waste'. *Journal of Economic Dynamics & Control*. 27: 157–80; Sundqvist, G. (2002). *The Bedrock of Opinion: Science, Technology and Society in the Siting of High Level Nuclear Waste*. (Kluwer, the Netherlands).

[32] Rechard, R. (2014). 'Site Selection and Regulatory Basis for the Yucca Mountain Disposal System'. *Reliability Engineering and System Safety*. 122: 731–50; Kraft, M. (2013). 'Nuclear Power and the Challenge of High Level Waste Disposal in the United States'. *Polity*. 45(2): 2–24; North, A. (2013). 'Getting U.S. High-Level Nuclear Waste into a Geological Repository'. *Risk Analysis*. 33(1): 1–14; Schimmoller, B. (2012). 'Zombieland and Nuclear Waste'. *Power Engineering*. 116(5): 18; Anon. (2012). 'Nuke Plants Put on Hold'. *New Scientist*. 18 Aug. 5; Alley, W. (2011). *Too Hot to Touch: The Problem of High Level Nuclear Waste*. (Cambridge UP, Cambridge). 19–20; Schaffer, M. (2011). 'Toward a Viable Nuclear Waste Disposal Program'. *Energy Policy*. 39: 1382–8; Anon. (2010). 'But Not in Yucca'. *New Scientist*. 6 Feb. 7; Vandenbosch, R. (2007). *Nuclear Waste Stalemate*. (Utah UP, Salt Lake). 50–96, 106–28, 139–66; Anon. (2006). 'Yucca Won't Blow'. *New Scientist*. 9 Sep. 6; Anon. (2002). 'Hotter in Nevada'. *New Scientist*. 23 Feb. 7; O'Hanlon, L. (2002). 'Yucca Nuclear Dump on Shaky Foundations'. *New Scientist*. 20 Apr. 11; Mason, B. (2002). 'Yucca Mountain Could Become Nuclear Volcano'. *New Scientist*. 24 Aug. 9.

[33] Gallardo, A. (2014). 'Geological Storage of Nuclear Waste: Insights Following the Fukushima Crisis'. *Energy Policy*. 73: 391–400; Birkholzer, J.

## 5. CONCLUSION

Nuclear waste is the only waste that will still be on the planet, and remain dangerous, for thousands of years into the future. It is the most hazardous of all waste streams. Despite a clear recognition of the risks that nuclear waste presents, the best that the international community has come up with is the prohibition of dumping nuclear waste into the oceans. After this, the guidance is to deal with nuclear waste via either reprocessing and/or deep geological storage. The problem with both of these options is that the first is hampered by considerations of economics and security, whilst the latter is limited by serious question marks over the feasibility of being able to safely hold such waste for perpetuity. Accordingly, a number of countries are backing away from either one or both options. The end result is that, as the problem continues to grow and it continues to be dealt with inadequately in short term holding facilities, there is no adequate answer on what to do in the longer term.

(2012). 'Geologic Disposal of High-Level Radioactive Waste: Status, Key Issues, and Trends'. *Annual Review of Environment and Resources*. 37: 79–106; Anon. (2010). 'Strategic Cooperation Sets the Scene for Geological Disposal of Nuclear Waste in Europe'. *Energy & Ecology*. March. 72; Strom, A. (2008). 'Site Descriptive Modelling During Characterization for a Geological Repository for Nuclear Waste in Sweden'. *Applied Geochemistry*. 23: 1747–60; Vandenbosch, R. (2007). *Nuclear Waste Stalemate*. (Utah UP, Salt Lake). 215–50.

# 10. Conclusion

The first goal of this book was to show the scale of the problem that modern waste has become, and what contextual certainties can be taken as given when working on this topic. What I found was that there are four relative certainties on waste. The first is the definition of what waste is – something relatively solid, that is unwanted and has no further use. The second certainty is that, as the disposal of waste increases, the trend is to find the cheapest way to dispose of it. If those seeking to dispose of the waste have scruples and wish to follow the law (as opposed to dumping it illegally), the trend, as it has been for the last 200 years, is to dispose of it where there will be least resistance, and thus, the lowest economic cost. Historically, this meant disposing of waste in either the commons, such as the ocean or rivers, or near communities without the economic capacity to object. Historically, this consideration of economic weakness was found within the poor regions of a town, city or country. Now, it is also found internationally as either pure waste or waste generated from recycling operations moves towards poor communities (and them towards it) at great speed. The reaction to this problem in areas where there are economic alternatives has been the creation of a new type of social movement, known as NIMBY. These movements are no longer the monopoly of the working class in wealthy countries. They have, however, a much greater uptake in countries where the immediate economic benefits of dealing in waste trump the longer term concerns of environmental and social impact.

This certainty that waste flows towards the location where it is economically cheapest to dispose of it, is radically compounded by the sheer scale of waste that is currently being produced. Although some waste streams have stopped because of technological change, most have both diversified and increased. Each of the new types of waste requires a greater degree of management than traditional types. The difficulties that this entails have been enhanced by increases in human numbers and per-capita rates of consumption for everything from consumer goods to more energy. No country has yet reached a point where its production of waste has begun to decline, although for a few high income countries, the rate of growth has slowed considerably in some waste streams. However, for the overwhelming majority of countries, the waste streams continue to increase in terms of both scale and complexity. These increases, both individually and collectively at the global level, represent one of the most obvious challenges for sustainability in the 21st century. The recognition of this challenge is the fourth certainty about waste – that no one thinks that waste is a good thing. More often than not, it is a mistake in economic, social and/or environmental terms. It is for these reasons that individuals, private enterprises, and policy makers at the local, regional and international levels want to reduce waste. No one is suggesting that more waste is a good thing. Rather, the clear consensus is that less waste should be created.

The good news is that there is a broadly agreed answer of the best way to confront the problem of waste. That answer is in the application of what is known as a 'waste hierarchy'. The two aims of the hierarchy are to generate as little waste as possible in the first place, and then to extract the maximum practical benefit of the waste that is still produced. In terms of sequence, the hierarchy suggests that, in priority, individuals, communities and policy makers should pursue prevention first; then reuse; recycle; other recovery (such as energy recovery); and then – as a last resort – final disposal.

In terms of the first part of the hierarchy, preventing waste from being generated initially, three options are available. First, the product can be prohibited. The problem here, especially when dealing with any waste that is not obviously hazardous, is that few legislators wish to actively prohibit consumer choice. Accordingly, a second method becomes available. This is that consumers may voluntarily decide to steer away from the particular stream of waste. The difficulty with this approach is that such voluntary action has never done more than slightly dent the growth, perhaps by up to 10 percent, in any particular waste stream. The third option is progressive pricing, whereby the disposal of waste generates a specific economic cost, thus making it unattractive to dump. If done correctly and there are adequate mechanisms to support recycling (so there is an alternative to dumping), this option may make reductions in a waste stream by between 20 and 30 percent. However, this is a finely balanced mechanism, as without adequate compliance, the problem of illegal dumping may quickly expand as waste producers seek to find ways to avoid the added economic costs of trying to make the production of waste unattractive. Finally, the producers of the product can be brought into the equation via ideals of extended producer responsibility, whereby those who initially design and manufacture the product also become responsible for dealing with the waste item. This last option is growing in popularity with over 400 schemes, covering everything from consumer batteries to motor vehicles, in operation across in both developed and developing countries. This approach of extended producer responsibility, when the responsibility is obligatory and linked to specific waste reduction targets, has the strongest possible potential out of any of the waste reduction mechanisms. However, many of these schemes are not obligatory, are not linked to specific targets and only cover a handful of products.

Despite all of these limits, it is clear that green consumerism, progressive pricing and extended producer responsibility have the most positive chances to slow the growth in the production of waste. However, even in the best case scenarios, they will not offset the overall growth in the amount and types of waste being generated. Nevertheless, they are the most promising weapons we have to slow the expansion of the amounts of waste being generated, before considerations of disposal come into play.

If waste cannot be avoided in the first place, the next best option, in terms of the hierarchy, is to recycle it. Recycling is the most common of environmental responses for many citizens and private enterprises. It is a well-recognised goal in domestic, regional and international law and policy, with roots going back hundreds of years. For most developed countries, the primary justification for recycling in the 21st century is environmental, but in all instances, the economic context provides a very solid foundation. This context of recycling is evident in all waste streams, with some waste streams showing great success at recycling, whilst in others, such as paper, glass, rubber and biowaste, there is limited uptake of recycling, and the markets remain sluggish, often only existing because of governmental support.

This situation is different to the markets involving the recycling of metals (from ships to electronic waste) and some plastics. In these instances, despite clear environmental benefits, the profit margins alone are driving the industries. The problem is that these markets, which are largely international, can be driven so aggressively that they may end up causing significant environmental and/or social damage to the people engaged in the recycling and their surrounding environment. This is especially so when the recycling industries operate in informal contexts, and the countries are willing to tolerate degrees of damage in the pursuit of the economic benefits of recycling that more developed countries are not willing to tolerate. Due to such risks, the international community has developed a series of rules and guidelines through the Basel Convention, which are designed to mitigate the worst impacts of these recycling trades. However, it is premature to say if the noble intentions of international law in this area to make recycling safer and more environmentally sound will win out against the profit motive in all recycling sectors, and the desire to keep economic costs of recycling as low as possible.

The last option in the recycling section is known as energy recovery. This option, where something is burnt and the energy generated from the combustion is captured, evolved out of the 1990s when it became argued that the careful burning of waste, to dispose of it and make energy from it at the same time, may be the best option for dealing with the problem, and benefiting from it simultaneously. It would appear now that, if high standards and the state of the art technology are utilised, this may be a good option to explore. Whether this is the 'best' thing to do will depend very much on a complex matrix of equations. In particular, it is not always clear that incineration of waste is the best thing to do with it, in terms of life cycle analysis. Second, just as with open burning, the evidence suggests that if controlled burning is done inefficiently, pollutants will result and these may cause environmental and/or social damage. This is currently a particularly polarised debate, and in a contemporary setting, is very much determined by what is burnt, to what standard and how exposed people are to the potential source.

The topic of energy recovery leads into the general possibility of waste incineration. The burning of waste in the open is a practice which is now uncommon in developed countries. This is not the trend in developing countries, where open burning of waste remains one of the most common methods to reduce the size of the waste pile and some of the nuisances associated with it. This ongoing practice is despite the very clear environmental and social risks of poor incineration.

Where there is greater uncertainty over the best way forward is with the industrialised incineration of waste. Controlled burning, via either municipal, industrialised or specialised incinerators, is a growth industry that began in the 19th century, and today is common in both the developed, and developing, world. However, for over 100 years this industry was the source of large amounts of dangerous air pollution, as the wastes that were burnt were either inappropriate to combust, or poorly burnt. When this became a clear problem on the high seas with ships specially designed to collect and burn waste on the open ocean, the international community eventually banned the practice. There is no such prohibition on land. Rather, on land, the problem has been confronted with the imposition of high technical standards for the operation of incineration plants and strict emission limits for specific pollutants. These standards appear to have mitigated some of the impacts associated with the earlier, more inefficient, forms of incineration. However, this appearance only exists for the countries which possess the high standards and the state of the art technology to achieve them. In most other places, the inefficient combustion of any type of waste, whether done in the open or in a controlled environment, presents an ongoing and direct environmental and social threat.

If waste could not be burnt and disposed of into the air, an alternative option was to dump it into the ocean. Whilst domestic restraints in this area began in the developed world in the early part of the 20th century, international restraints did not emerge until the 1970s. The international restraints, as found through the London Dumping Convention, were the subject of intense debate for over a decade, as the regime only prohibited the dumping of the most hazardous of substances. Accordingly, the dumping of most types of waste into the ocean was the practice, not the exception. This approach was completely turned around in 1996 when a Protocol to that Convention reversed the assumption that it was safe to dump anything into the ocean, and only a few specific waste streams could be deposited into the seas. At this point, many of the practices that were previously acceptable, became unacceptable. This was particularly obvious with regard to the dumping of either industrial or nuclear wastes. In addition, other wastes which were barely regulated, such as sewage sludge or dredgings, have become much more regulated, buttressed with a presumption against dumping, unless it is clearly safe to do so. However, despite the strong progress in this area, it is clear that some of the most emergent problems facing the ocean in terms of waste at the moment, like plastic, cannot be solved by such regimes, as the source of the problem is on land. Accordingly, it is only if this problem is tackled, via methods such as producer responsibility, recycling and more effective waste management approaches involving collection, stopping litter and effective disposal on land, will the difficulty of this waste stream going into the oceans be solved.

The disposal of waste into the land is the most common form of waste disposal on Earth. Humanity began disposing of waste into the earth around 2000 BCE. The difference today is that the amounts deposited are infinitely greater as are the types and sources of waste being put into the land. The waste being put into the land today is often part of a global culture, and may have been generated on the other side of the planet to begin with. Since the 19th century, to mitigate against the risks of dumping waste into the land, successive generations have developed rules and policies to control and regulate this disposal method. These rules have grown from zoning requirements to detailed prescription of how to manage these sites, from what can be deposited in dumps and the pollution protection mechanisms to stop leachate. In the 21st century, these rules and principles have become well adopted and practised in most developed countries. The same cannot be said for many developing countries. This is a particular problem, for many of these countries are now dealing with their own rapidly expanding waste streams, as well as large amounts of imported material which is either intended to be disposed of as waste, or, via poor recycling standards, ends up as waste to be deposited into landfill.

This problem, of largely unregulated landfill in the developing world, is complicated further by the possibilities of international trade in this area. As a broad rule, international law facilitates, and does not prohibit, the trade in waste for final disposal, if the importing countries clearly consent to the waste entering into their country, and can assert that they can safely deal with it. Although the practice is becoming clear that, where a waste is clearly hazardous, it should only be shipped from the developing world to the developed, this is by no means absolute, and hundreds of thousands of tons of waste which is in the hinterland between hazardous and non-hazardous continues to progress to the poorer parts of the world. In theory, the countries that consent to these imports know what they are doing. In the political context that these countries make such decisions, and the subsequent ecological and/or social costs, this is often far from obvious.

The last two topics that this book looked at were what I consider to be the two most pressing waste streams. The first is human waste, and the second is nuclear. I focused on human waste as this particular type of waste causes more deaths than all of the other waste streams put together. Human waste is the most personal, and lethal, of all waste streams. It is also the first type of waste that humanity began to manage, some 5,000 years ago. However, human populations have both expanded and urbanised, and the risks of disease have increased, and the ability to adequately manage human waste has not always been clear. In the 19th century, the first generation of problems involved getting the waste out of the cities. Second generation problems involved getting the waste suitably treated and neutralised to the greatest possible degree, and then finally, as a third generation problem, effectively and safely disposing of the residual waste, which is typically in the form of sludge.

This need, to remove human waste safely from places where it can cause disease, and then adequately deal with it, is a well-recognised international goal. However, success in meeting this goal is not the experience of all countries. Whilst most developed countries are adequately confronting this problem, a number of developing countries are still struggling with the first parts of this equation, trying to ensure elementary sanitation and basic evacuation procedures for this waste. It is only for the countries which have mastered the first generation question of getting the human waste away from settlements that the issues of the final disposal of human waste come to the fore. The options for the final disposal are into freshwater systems and/or the ocean; or disposal via either incineration, placement into landfill or recycling. Regarding the water option, whilst the international community has come to regulate, not prohibit, the dumping of such wastes into the ocean with a relative degree of success, the same cannot be said for freshwater systems. Here, despite some broad policy goals at the global level, the success of keeping human waste out of freshwater systems has ranged from very good in many developed countries to very poor in many developing countries and associated regions. In respect of the other options of disposal via incineration, landfill or recycling, each approach has its benefits, and its limits. It is possible that the last option, where biosolids are fed back into a well-managed biotic system, is the most advantageous use of the resource. However, this is only utilised by a few countries, and even then, within a tightly regulated context.

Nuclear waste is the only waste that will still be on the planet, and remain dangerous, for tens of thousands of years into the future. It is the most hazardous of all waste streams. Despite a clear recognition of the risks that nuclear waste represents, the best that the international community has come up with is the prohibition of dumping nuclear waste into the oceans. After this, the guidance is to deal with nuclear waste via either reprocessing and/or deep geological storage. The problem with both of these options is that the first is hampered by considerations of economics and security, whilst the latter is limited by serious question marks over the feasibility of being able to safely hold such waste for perpetuity. Accordingly, a number of countries are backing away from either one or both options. The simple conclusion with this waste stream is that no one really knows what to do with it. Unlike all the other waste streams where there are at least agreed technical answers on what is the best way forward, and many countries are implementing these, the same is not true for nuclear waste. The end result is that the waste stream continues to grow and it continues to be dealt with inadequately in short term holding facilities, and there is no adequate answer on what to do in the longer term.

After all of these words, here is the conclusion of this work. Humanity has a long history of dealing with waste. However, we have never had to deal with so much, of so many varieties and of so many sources. We have a problem on our hands. This is not news. We have known of the need to adequately deal with waste streams for over 100 years. Today, we work on the principles that we should seek to prevent waste at source, and if this is not possible, we should recycle it. Only if these two options are not possible, should we dispose of it. However, although we have the economic and policy tools to reduce waste streams via both pressure on consumers and producers, these are only adequately applied in a few very select areas. Similar restraints exist upon recycling, where the economics of the equation either dampen or enflame the operations. And as above, success in this area is restricted to only a few waste streams. With regard to energy recovery operations, it is too early to say when this should, or should not, be the preferred method of disposal. When it clearly should not be considered an adequate waste disposal method, as with all incineration options, is when burning is done inefficiently and/or the wrong wastes are burnt. Uncontrolled and/or inefficient burning should be prohibited everywhere, but it is not. In many developing countries, it is barely regulated and is often standard practice. Unfortunately, the same conclusion – that standards exist to control the worst parts of waste disposal, but they are commonly not applied in the developing world – also exists for the problems of disposal of waste into the land and of human waste. In this regard, the question of whether waste is being adequately dealt with or not often comes to depend very much upon the economic status of the country dealing with it. With nuclear waste, it is not a question of which country ends up holding it. Here, the issue is much simpler. It is simply not clear that there is an adequate way to hold this type of waste, which continues to proliferate.

# Index